T0281486

Dynamic Isolation Technologies in Negative Pressure Isolation Wards

Zhonglin Xu · Bin Zhou

Dynamic Isolation Technologies in Negative Pressure Isolation Wards

 Springer

Zhonglin Xu
China Academy of Building Research
Beijing
China

Bin Zhou
Nanjing Tech University
Nanjing
China

ISBN 978-981-10-9740-9   ISBN 978-981-10-2923-3   (eBook)
DOI 10.1007/978-981-10-2923-3

Printed on acid-free paper

This Springer imprint is published by Springer Nature
The registered company is Springer Nature Singapore Pte Ltd.
The registered company address is: 152 Beach Road, #22-06/08 Gateway East, Singapore 189721, Singapore

Preface

After the epidemic of the infectious disease SARS occurred in 2003, a simulated negative pressure ward was set up in the Institute of HVAC at China Academy of Building Research. The series of experiments were carried out by me and a research group from the institute. Ten serial research reports and one monograph were published. The proposal of one concept and its corresponding technical measures was completed. The first construction standard in China for negative pressure ward was compiled in Beijing, which has been referred and adopted by other local standards.

Nowadays with frequent occurrence of airborne infectious diseases, the new manuscript is prepared by me based on the above-mentioned research reports and monograph. It is hoped that the effective measures proved by research could be introduced to more people. Assoc. Prof. Dr. Zhou Bin was then invited to translate the English version of this book based on the Chinese version. His brilliant work was greatly appreciated.

For further requirement, please do not hesitate to contact me by ph_ph@163.com.

Beijing, China
April 2016

Zhonglin Xu
From Xu Zhonglin Studio
Institute of Building Environment and Energy
China Academy of Building Research

Contents

Chapter 1
Importance of Negative Pressure Wards

1.1 The Disaster at the Beginning of the Century

At the early beginning of the 21st century, the disaster with airborne infectious diseases appeared in China.

On November 6th in 2002, the first patient with Severe Acute Respiratory Syndrome, or SARS, was diagnosed in the city Foshan of Guangdong Province. On February 11th in 2003, the first formal report about the epidemic was delivered to WHO. At that time, 305 people have been infected and 5 people were dead because of this disease. Till August 7th in 2003, the epidemic has been dispersed into 34 countries and regions within the short period of 6 months. The number of the cumulative suspected cases reached 8347. 916 people were dead. In that period, the freedom of travel for most people was restricted. The activity of economy and trade was severely affected. Take the Far East as an example, the estimated economic loss reached 30 billion dollars [1].

SARS will not affect only patients, but also the medical personnel in ward and related area. The infectious rate of medical personnel in Beijing and Hong Kong reached or exceeded 20%, while that in Taiwan reached as high as 30%. 20% of the infected people in the globe were medical personnel.

The reason for this phenomenon is that little attention has been paid on infectious disease in the globe during the past decades. There exist the following misunderstandings:

(1) It is believed that the development of economy is helpful to the natural decay of infectious disease. The main treats for human health are cardiovascular disease, tumor and diabetes.
(2) Infectious disease can be overcome by microbiology technology.
(3) The fact of long-period existence of infectious disease and the trend to become the first killer for human being are overlooked.

© Springer Nature Singapore Pte Ltd. 2017
Z. Xu and B. Zhou, *Dynamic Isolation Technologies in Negative Pressure Isolation Wards*, DOI 10.1007/978-981-10-2923-3_1

With the above-mentioned misunderstandings, the investment on the infection prevention projects of epidemic disease for public health reduced sharply. The number of hospitals for infectious diseases in the whole nation decreased. The instruments inside were odd. There were no real isolation wards as infectious wards. Even for abroad such as U.S.A., most of the hospitals for infectious diseases were closed in 1950s.

In ordinary ward, indoor air distribution is not good. Polluted air cannot be exhausted outdoors quickly and efficiently at all. Therefore, the safety of medical personnel cannot be guaranteed.

Because the technique to effectively isolate airborne infection was rare and there was no existing standardized isolation ward, only some general provisions can be issued at first. Some temporary emergence measures were enforced or suggested.

In April of 2003, WHO provided principled recommendations on SARS ward in the revised "*Hospital Infection Control Guidance for Severe Acute Respiratory Syndrome*", including negative pressure, single room with toilet, independent air supply or exhaust.

On April 30th in 2003, the general office at Ministry of Construction of the People's Republic of China and Ministry of Health of the People's Republic of China issued an "emergency notice". It required that central air conditioning system is prohibited in the places where enrollment, isolation and inspection of SARS cases were taken place and where SARS patients were diagnosed. This was meant to prevent the transmission routine of SARS virus. On May 5th in 2003, Ministry of Health of the People's Republic of China issued "*Guidelines for Infection Control of SARS in Hospital (Trial)*". It re-emphasized that both fever clinics and isolation observation room should be set relative independently in easily isolated places inside the hospital. Special ward should be set in hospitals where SARS infected patients are received and treated. Natural ventilation for air convection should be guaranteed between indoors and outdoors. Ventilation equipment (such as fan) must be installed for these places where the performance of natural ventilation is poor. Central air conditioning system is prohibited. In the same month, "*Design Highlights for Hospital Buildings Receiving SARS Infected Patients*" were issued, which provided some fundamental general provisions.

1.2   Severity of Airborne Infection

The infectious diseases are mainly transmitted through two ways, including contact transmission and airborne transmission. It should be noted that except for the well-known respiratory transmission pathway, disease can also be infectious through contact transmission. In this case, microbial aerosol may deposit on surfaces including body, hand and other accessible surfaces, if sterilization on air is not performed thoroughly. Then infection may occur when these surfaces contact the susceptible sites of infection. There is a logion from Dr. Hinds that "Viruses are transmitted by direct contact or by inhalation of aerosolized viruses" [2].

There are mainly 41 kinds of infectious diseases, among which 14 kinds are airborne. These airborne diseases rank the first in various specific transmission pathways. In the whole globe, the infectious diseases resulting from respiratory infection by microbial aerosol occupy about 20% of the total diseases. In China, the proportion of respiratory infection occupies 23.3 ∼ 42.1% of nosocomial infection, which ranks the first position in various specific transmission routes.

There are about 30 kinds of pathogens causing the respiratory infection through airborne transmission.

For bacteria, it includes *Streptococcus Pneumoniae, Escherichia Coli, Pseudomonas Aeruginosa, Klebsiella, Serratia Marcescens, Salmonella* sp., *Legionella spp., Mycobacterium Tuberculosis, Staphylococcus Aureus, Enterococcus*, etc.

For fungi, it includes Fumagillin, Rhizopin, Trichothecin, Candida Albicans, Histoplasma Capsulatum, etc.

For virus, it includes Coronavirus, influenza virus, measles virus, varicella-zoster virus, mumps virus, variola virus, Swine vesicular disease virus, Hemorrhagic fever virus, and Coxsackievirus.

For Rickettsia, it includes Q Fever.

For others, it also includes Mycoplasma and Chlamydia.

The most serious case for respiratory infection should be the pulmonary tuberculosis. In 1994, the Centers for Disease Control and Prevention (CDC) in U.S.A. issued *"Guidelines for Preventing the Transmission of Mycobacterium tuberculosis in Healthcare Facilities"*. It pointed out that the new case of tuberculosis infection would become the active tuberculosis quickly. The death rate related to the sudden occurrence of this disease is quite high, which is between 43 and 93%. Moreover, the period between diagnosis and death is quite short. The median of the interval period is from 4 to 16 weeks.

The number of tuberculosis patients from China takes the second place in the world. The number of death because of pulmonary tuberculosis reaches as more as 130 thousands in 2004, which greatly exceeds the total number of death resulting from other infectious diseases.

The severity of the respiratory infection by airborne transmission includes the following aspects:

(1) explosive epidemic—In a short time, a large amount of people will be infected. Take the incidence of Legionella infection happened in U.S.A. in 1976 as an example, the mortality of the people caused by this disease was 29.
(2) large infection area—The disease can be epidemic in all the country or all the world, which can be evidenced by historical pandemic influenza.
(3) extremely low infection dose—The dose of infection is much lower than that through other transmission routes. People will be easily infected if only one Q fever deposits onto the respiratory tract. People will be infected if more than 0.1 billion REPECs were ingested, while infection will occur if 10 to 50 REPECs were inhaled. The median infective dose of respiratory by adenovirus is only a half of that by tissue culture. Experiment also shows that even though the inhaled quantity is larger than the minimum infectious dose when the

inhalation period is prolonged, there is still no infectious effect. Therefore, in terms of the dose for infection control, except for "the concentration of airborne droplet nuclei to prevent transmission and reduce infection" required in *"Guidelines for Preventing the Transmission of Mycobacterium tuberculosis in Healthcare Facilities"*, special attention should also be paid on the virus dose which has the infection effect. It should also be noted that the initial exposure airborne virus concentration plays the leading role.

4) cross infection—Although *Salmonella* sp. is the bacteria for infections of the digestive tract symptoms, respiratory infection case appears in the pediatric surgery ward. Venezuelan equine encephalitis virus and yellow fever virus are arboviruses, but they can cause respiratory infection. Other people can be infected by hepatitis virus and syphilis through microbial aerosol generated from dental drill with high rotation speed [2].

Based on the investigation of infectious disease in a neonatal ward lasting for 11 months, it was found that the outbreaks of seven Gastroenteritis cases by Salmonella typhimurium (Bacteriophage T2) were not from contaminated food, but from collected dust contaminated by this bacteria in a vacuum cleaner. In ten months after the occurrence of the last infection case, Salmonella typhimurium can still be separated from dust.

In the past, it was believed that the Measles was caused by direct contact infection with airborne droplet nuclei. However, during an outbreak of Measles in a pediatric clinic in a hospital located in State of Michigan in U.S.A., it was found that three out of four infected children had no history of direct contact. It was the cough from infected children that may cause airborne infection. It could emit 144 infectious virus particles per minute during cough [3].

There is also evidence of infection through airborne transmission for Hepatitis B Virus which spreads mainly through digestive tract infection. Therefore it is reasonable to consider that it is more likely to cause environmental contamination for Hepatitis B Virus through saliva and airborne droplet nuclei than that through contact by hands. Table 1.1 illustrates the positive detection rate of virus in air sample [3].

Airborne transmission mainly depends on aerosol and airborne droplet nuclei. During the conversation, cough and sneeze of patient, not only aerosol but also spittle and droplet nuclei after evaporation can be generated.

Table 1.1 Detection rate of HBsAg in air sample from public place

Sampling position	Sample number	Positive sample number	Positive rate (%)	Sampling position	Sample number	Positive sample number	Positive rate (%)
Hospital clinic department	10	3	30	Waiting room	8	2	25
Karaoke room	8	2	25	Outskirts	5	0	0

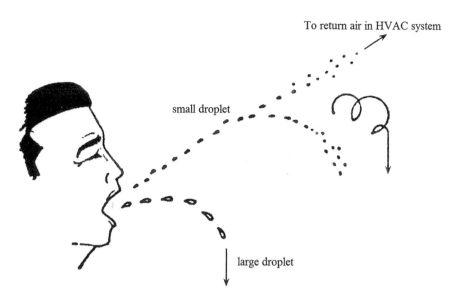

Fig. 1.1 Droplet nuclei and aerosol generated during sneeze

Figure 1.1 and 1.2 shows the situation of sneeze. Figure 1.3 illustrates the number of particles released during sneeze, which can reach as more as 300 thousands. The number of aerosol released from respiration of patients depends on the expiratory velocity. The number in common case is not too much. Although

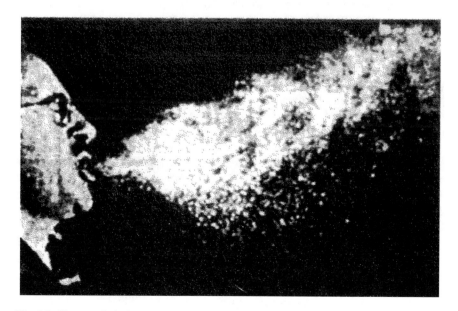

Fig. 1.2 Photograph during sneeze

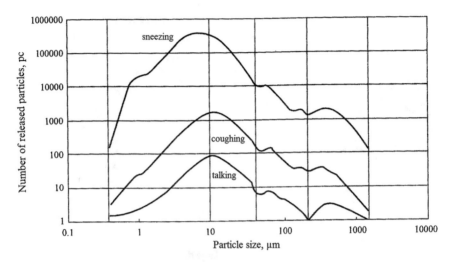

Fig. 1.3 Particle size distribution of generated aerosol released during sneeze

only thirteen infectious droplet nuclei could be released from a tuberculosis patient per hour, twenty seven people could be infected within four weeks [2].

The airborne droplet aerosol is spread in the following way:

Spittle will migrate under the influence of airflow, gravitational force and drag force. Therefore, the distance of movement is very small, which is less than 0.9 m [4].

However, the influence of gravitational force on aerosol is very small. Aerosol will move under the influence of airflow mainly. When it is near the return air outlet, it will be sucked towards the outlet. Therefore, aerosol can disperse to further place.

Therefore, the transmission characteristics between spittle (large droplet) and aerosol (small droplet and droplet nuclei) are distinct. But most of large spittle will become droplet nuclei eventually, i.e., aerosol. As for the Coronavirus, it exists mainly in secretion and droplet. So except for the transmission from direct contact, the route of airborne transmission is also very important.

The outbreak of Severe Acute Respiratory Syndrome (SARS) at Amoy Gardens in Hong Kong proved that SARS Coronavirus can spread through the route of airborne transmission. According to the investigation report of SARS outbreak at Amoy Gardens, which was issued by Department of Health in Hong Kong on April 17th in 2003, and based on test with oil droplet aerosol by Hong Kong Polytechnic University, it was found that droplet will move upwards with the airflow of "fume", and it will also disperse transversely, which was influenced by the exhaust and return air from the exhaust fan at external window inside the bathroom at each floor. It took only several seconds for this kind of flow to reach the upper ceiling.

Therefore, it is not acceptable to reach the conclusion that it is safe to stay three feet away, which is based on the fact that the transmission distance of spittle is less than 1 m [5].

Whether pathogen can be spread through spittle or aerosol (droplet nuclei) depends on the degree of dependency on the water content and the nutrition provided by the carrier. If the degree of dependency is large, aerosol with small diameter may not be able to provide enough water content and nutrition for its survival. In this case, the movement of aerosol cannot be the transmission route of infection.

Previous studies have shown that the maximum diameter of droplet and spittle can reach 100 μm. It is reasonable to consider the size to be continuous. After evaporation, they will become aerosols with smaller diameter, i.e., the so-called airborne droplet nuclei. When they enter into the flow stream of recirculation air, the complexity and risk of transmission for Coronavirus will be increased. Therefore, it is necessary to control the exhaust air in the long distance, such as the influence of exhaust air from sewerage at Amoy Gardens in Hong Kong. It is also necessary to isolate in short distance or take primary isolation measures, which can be proved by the reality that most medical personnel close to SARS infected patients got infected.

It is usually beyond the expectation that microbial aerosol can also be generated from the flush toilet [2]. American scholar named Wallis reported the research result about the aerosolization of poliovirus during the usage of the flush toilet in 1985. Air was sampled during the flush of water after virus was injected into the urinal manually. Results were shown in Table 1.2.

Table 1.2 Aerosolization situation of poliovirus during the usage of the flush toilet

No.	Dose of injected virus	Number of collected virus with microporous membrane
1	3.8×10^8	1545
2	4.3×10^9	1383
3	6.1×10^9	775
4	4.8×10^9	455
5	5.0×10^9	1270
6	2.4×10^9	563
7	3.0×10^9	1600
8	3.0×10^9	704

Table 1.3 Generation of aerosol during flush of water on the fresh faeces containing virus

Mass of faeces (g)	Total amount of virus (PFU)	Virus in water from urinal (PFU/mL)	Recycled virus (PFU)
24	2.4×10^7	2040	65
50	4.5×10^7	480	6
64	4.8×10^7	180	0
22	5.9×10^7	28	0

Sampling results with the flush of water on the fresh faeces containing virus are shown in Table 1.3.

In short, the following environmental factors may be responsible for the spread of aerosol and spittle:

(1) The space is relative small and closed, where infectious particle will accumulate.
(2) There are no enough local or overall ventilation rates. The infectious particles cannot be diluted, whose concentration (so-called infection dose) is high enough to be epidemic.
(3) For recirculation air containing infectious particles, the risk of infection increases.
(4) Under the effect of pressure difference, infectious particles can penetrate from one side to another. They may migrate from one side to another with the influence of airflow. These provide the opportunity to increase concentration.

Therefore, all the measures taken to control the transmission of aerosol and spittle are reducing the concentration.

Reference [6] provides the theoretical analysis for the transmission of aerosol and spittle.

1.3   Requirement for Negative Pressure Ward

There are four classification levels of infectious isolation wards, according to the contagious strength of the disease from patients [7].

Class 1: Contact isolation, such as for Hepatitis A patient.
Class 2: Droplet isolation. The difference from Class 1 is the operational procedure.
Class 3: Air isolation, such as for Tuberculosis patient.
Class 4: Sealing isolation, such as for patient infected by Ebola virus fever, Ebola hemorrhagic fever, SARS, Staphylococcus aureus. This kind of isolation has the dual functions of contact isolation and air isolation.

During the early stage for the outbreak of SARS, the most urgent task was to relocate these confirmed and suspected patients into isolation ward as soon as possible, which was meant to avoid cross infection. However, the consciousness of isolation among medical personnel was weakened these years. They were not wary and showed contempt for infection with airborne transmission. This caused the weakened situation for the scientific research in the field of health care with infectious disease. The scientific study on construction of isolation ward was totally vacant.

In order to face the emergent event, common wards were reconstructed into isolation wards in mainland China, Taiwan, Hong Kong and other regions. In other places, only some simple measures were taken.

Fig. 1.4 One of the most
simplest isolation forms in
emergency situation

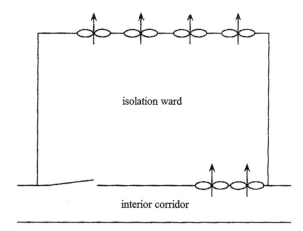

isolation ward

interior corridor

For example, in order to meet the urgent needs during the early stage of the outbreak of SARS in March, 2003, window-type exhaust fans were added in renovated ward in some public hospitals in Hong Kong. Air from corridor could enter into the ward, and was then discharged outdoors after passing through the sickbed.

The same method was adopted in Guangzhou Eighth People's Hospital during the epidemic of SARS in 2003. Five to six exhaust fans were installed on exterior window of the ward. Two supply fans were set on interior wall connecting the corridor and the ward. In this way, indoor air could not flow reversely towards the corridor, and then went outside of the ward area. The effect of isolation with airflow was achieved, which is shown in Fig. 1.4.

The application showed that there was no cross infection inside ward area. However, because of the large velocity, patients felt cold even when the quilts were used during the early spring period.

In fact, it is not necessary to install five to six exhaust fans, because the room is not large. Additionally, the supply fan is also unnecessary. It is fine if one louver were installed above the door.

Of course, the influence of exhaust air on atmosphere was not considered. This is only one of the simplest isolation forms in emergency situation. At the late stage of SARS epidemic, it was required that HEPA filter must be installed in exhaust air pipeline during the formal design process in Hong Kong.

Moreover, staffs including operators, laboratory technicians and medical personnel will directly face the recipients, i.e., the pollution source. They may be the first victims. As mentioned before, about 20% of infected people were medical personnel during the epidemic of SARS in 2003. In Hong Kong, the proportion reached 22%.

In order to protect operators (medical personnel) from infection, the measures taken was termed as Primary Isolation.

Setting of exhaust hood near the head of sickbed is one of the isolation methods with physical barriers.

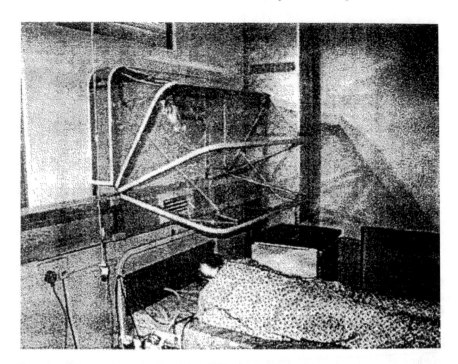

Fig. 1.5 Movable exhaust hood near the head of the sickbed

The exhaust hood near the head of sickbed should be flexible. During the SARS period in Hong Kong, one movable exhaust hood was tested [8], which is shown in Fig. 1.5. The transparent hood was used to isolate the exhaled aerosol. Air was exhausted through the exhaust air outlet near the head of the sickbed or through the exhaust fan on the hood. However, practice showed that this kind of isolation was not welcomed, because it limited the movement of patients including lying down and getting up (shown in Fig. 1.6) and it also influenced the draft sensation of patients. Therefore, it was not promoted during the SARS epidemic period.

In brief, according to the practical cases of isolation wards investigated by author at home and abroad at that time, few of isolation wards of this kind can meet the fundamental requirements for the control of airborne infection. The fundamental requirements include the following aspects:

(1) To protect other patients and medical personnel from infection.
(2) To protect outdoor environment from infection.
(3) To prevent cross infection between patients.

Isolation wards which meet the above requirements are negative isolation wards, which are different from the protective isolation wards with positive pressure.

In 2009, the A/H1N1 influenza virus outbroke in the world. With the conditions of continuous appearance of novel infectious disease and re-appearance of

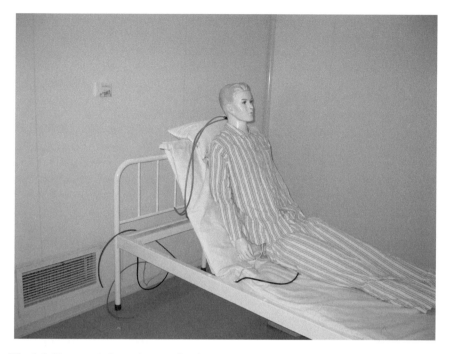

Fig. 1.6 Photograph for getting up of patient

existing/old infectious disease, the local standard DB 11/663-2009 "*Essential construction requirements of negative pressure isolation ward*" was issued in Beijing, which plays an important role in guiding the design of isolation ward in Beijing and other places and controlling the disperse of infectious disease.

References

1. WHO, Acute Respiratory Syndrome (SARS): status of the outbreak and lessions for the immediate future (2003)
2. X. Yu, *Modern Air Microbiology* (People's Military Medical Press, Beijing, 2002)
3. F. Che, *Principle and Application of Air Microbiology* (Science Press, Beijing, 2004)
4. B. Zhao, Z. Zhang, X. Li, Numerical study of indoor spittle transportation. J. HV&AC **33**, 34–36 (2003)
5. W.J. Kowalski, W. Bahnfleth, Airborne respiratory diseases and mechanical systems for control of microbes. Heating Pip. Air Cond. **70**(7), 34–48 (1998)
6. Z. Xu, *Design Principle of Isolation Ward* (Science Press, Beijing, 2006), pp. 5–18
7. H. Huiskamp, Infection control measures in hospitals in Netherlands. in *Proceedings of the 6th China International (Shanghai) Academic Forum & Expo on Cleanroom Technology*, 2003
8. F. Chan, V. Cheung, Y. Li, A. Wong, R. Yau, L. Yang, Air distribution design in a SARS ward with multiple beds. Build. Energy Environ. **23**(1), 21–33 (2004)

Chapter 2
Three Misunderstandings for Design of Negative Pressure Ward

With the severe situation for the appearance of SARS and the fear for the resultant consequence, at the early stage of the outbreak event wards were reconstructed with a simple way. Newly constructed isolation wards were designed according to related literatures issued by CDC in 1994, and the corresponding requirements were elevated blindly. In literatures, the technical measures related to negative pressure isolation ward was inclined to adopt high negative pressure, air-tight door and all fresh air, where it was considered safe only to increase the negative pressure as high as possible, to install air-tight door without infiltration air and to provide all fresh air.

Therefore, three misunderstandings were taken during the design of negative pressure isolation ward at one time, which included high negative pressure, air-tight door and all fresh air. This will cause large amount of waste for investment and energy. Sometimes it even causes severe accident.

2.1 About High Negative Pressure

2.1.1 Effect of Pressure Difference

The so-called high negative pressure means that negative pressure should be kept between isolation ward and corridor or adjacent room. The negative pressure drop should be as high as dozens of Pascal. It is meant to prevent the release of hazardous indoor air from flowing outwards.

Is it really effective to prevent the leakage of airflow outwards? Is it better to adopt higher pressure drop?

The positive pressure inside the ward could be utilized to prevent the invasion of infectious air through the gap into the ward, which is shown in Fig. 2.1. The negative pressure inside the ward could be applied to prevent the release of infectious air through the gap towards outside of the ward, which is shown in Fig. 2.2.

© Springer Nature Singapore Pte Ltd. 2017
Z. Xu and B. Zhou, *Dynamic Isolation Technologies in Negative Pressure Isolation Wards*, DOI 10.1007/978-981-10-2923-3_2

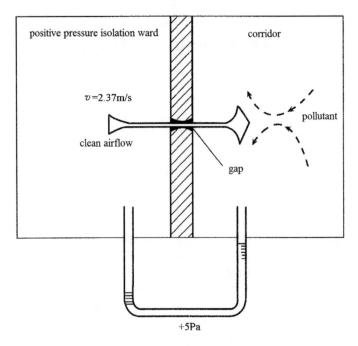

Fig. 2.1 Schematic diagram for prevention of infectious air by positive pressure

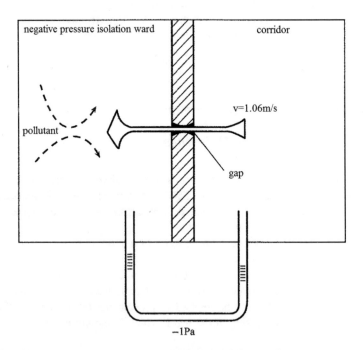

Fig. 2.2 Schematic diagram for prevention of infectious air by negative pressure

The effect of pressure difference on the prevention of leakage though the gap appears only when all the openings between the isolation ward and the adjacent rooms are closed.

Therefore, the pressure difference is only the main measures to realize static isolation.

This characteristic of the pressure difference is based on the time characteristic of the pressure difference.

If the door is open, the pressure difference disappears. In total, the pressure of air at both sides reaches equilibrium.

Figure 2.3 illustrates the experimental result performed by a Japanese scholar [1]. Results showed that for a room with the pressure difference of −15 Pa from the outside, the pressure difference would decrease to 0 within 1 s when the inwardly opened door was open. For outwardly opening door and sliding door were open, the period for this decrease of pressure difference could be prolonged to 2 s. All of them illustrate that there is no effect of pressure difference when door is open.

It was also pointed out in the Handbook of ASHRAE in 1991 that the original pressure difference between two regions would reduce to zero instantaneously when the door or the closed opening between two regions was open [2].

Therefore, it was confirmed in "*Guidelines for Preventing the Transmission of Mycobacterium tuberculosis in Healthcare Facilities*" issued by the Centers for Disease Control and Prevention (CDC) in U.S.A. in 1994 that "the crucial problem is to keep the door and the window between isolation ward and other region closed, except the case when people go inward or outward". It should be noted that here it only mention the state of close, not the sealing condition, which will be explained later.

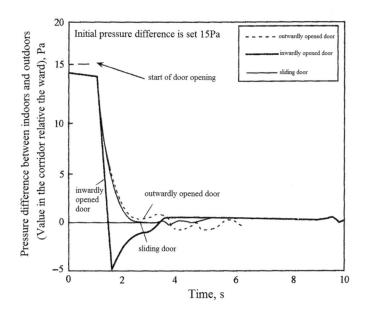

Fig. 2.3 Variation of pressure difference with time during the opening of door

So the main purposed of the pressure drop is only limited to the static state of the closed openings. During the instantaneous moment for dynamic opening of door, the pressure drop is converted into the kinetic energy of airflow through the opening. The magnitude of airflow velocity reflects the ability to prevent the entrance of pollutant, which is not dependent on the magnitude of original pressure difference. For the given flow rate with compensation of pressure difference, the value of airflow velocity is fixed.

In the past, there is an incomplete understanding that for isolation ward, isolation cleanroom and biosafety laboratory, the principle of isolation mainly depended on the effect of gradient of negative pressure (i.e., the negative pressure difference). The negative pressure was considered as the sole measures to prevent the release of pollutant outwards for isolation ward and biosafety laboratory. In the early study, we have investigated and found the instantaneous influence of opening and closing of doors and the entrance and exit of people on the counteracting effect by the pressure difference [3]. Further detailed investigation for the principle of isolation ward was carried out after the outbreak of SARS [4].

2.1.2   Ability to Control Pollution Dispersion by Pressure Difference

In the above-mentioned ASHRAE Handbook, positive and negative pressure could also be considered as the measures to resist other factors. Because of the opening of doors, movement of workers and patients, temperature difference, and the stack effect exaggerated by straight pipeline, elevator shaft and vertical ventilation shaft, it is difficult to control airflow reasonably between rooms. When some factor becomes larger than the actual controlled range, the influence of these factors could be minimized through the modification during the design of positive and negative pressure values in some rooms or regions.

It has been pointed out by author that this incomplete understanding is caused because that they do not understand the invasion or release of pollution and they do not know that the pressure is not the sole factor [3]. This can be analyzed from two aspects as follows.

(1) On the one hand, the air velocity through the door openings by the pressure difference is very small, which could not prevent the outward leakage or invasion of pollutant after the door is open. The flow rate of leakage through gap could be derived further by the air velocity through the gap, i.e.,

$$Q = 3600 \cdot F \cdot v = 3600 \mu F \sqrt{\frac{2\Delta P}{\rho}} \qquad (2.1)$$

where μ is the flow coefficient. It is usually between 0.3 and 0.5, so we can use 0.4. F is the area of the gap, m^2. ΔP is the pressure difference, Pa. ρ is the air density, which can be assumed 1.2 kg/m^3.

Table 2.1 illustrates the flow rate of leakage for a room with area 15 m^2 under the condition of different pressure drops.

In this room, the dimension of the gap on the air-tight door is 6 m × 0.0005 m, while that on the non air-tight door is 6 m × 0.005 m. The dimension of the gap between the air-tight window and the delivery window is 8 m × 0.0005 m. The size of the gap on the wooden partition board is 40 m × 0.0001 m.

It is shown in Table 2.1 that when the non air-tight door is open, because of the high pressure difference ($\Delta P = -30$ Pa), the air flow rate of leakage in the whole room is converted into the flow rate of entrance air through the gap which is only 0.101 m^3/s. When air-tight door is open, it becomes 0.026 m^3/s, which corresponds to the average air velocity through door opening 0.11 m/s. It is larger than the convection velocity resulting from the temperature difference 0.1 °C by 0.035 m/s, which is very small and will be mentioned later. If the negative pressure difference is −15 Pa, the resultant velocity will be much smaller than the convection velocity.

Therefore, it is essentially a subjective judgment to consider that "for a room with negative pressure when doors are closed, the flow rate during the opening of door mainly depends on the magnitude of negative pressure difference [5]".

Because of the small flow rate, the effect to prevent pollution dispersion by pressure difference is limited. Therefore, it is also pointed in Ref. [6] by foreign researchers that negative pressure should be kept inside isolation room, but the

Table 2.1 The flow rate of leakage under different pressure drops

Pressure difference	Air velocity through gap	Flow rate of leakage with non air-tight door	Flow rate of leakage with air-tight door	Air velocity through door when non air-tight door is open
ΔP, Pa	$\bar{v}$, m^3/s	Q, m^3/s	Q, m^3/s	$\bar{v}$, m^3/s
1	0.52	0.019	0.006	0.021
2	0.74	0.026	0.007	0.029
3	0.90	0.033	0.009	0.037
4	1.05	0.037	0.010	0.041
8	1.48	0.053	0.014	0.059
10	1.64	0.058	0.015	0.064
15	2.01	0.072	0.019	0.081
20	2.33	0.083	0.022	0.092
25	2.60	0.092	0.024	0.102
30	2.85	0.101	0.026	0.112
35	3.08	0.110	0.028	0.122
40	3.29	0.117	0.030	0.13
45	3.49	0.124	0.032	0.128
50	3.68	0.131	0.035	0.146

magnitude of negative pressure is not important (the same to positive pressure). This is proved in detail by Tables 2.2 and 2.3. For the isolation room with negative pressure, when it becomes positive pressure with magnitude as small as 0.001 Pa (close to 0 Pa), the leakage rate of microbial particles reaches 1.3×10^4CFU/year, and vice versa.

Table 2.4 shows the experimental data when atmospheric dust was applied to study the isolation room by Chinese researchers [7].

In the experiment, the following steps were adopted. At first, full fresh air without air filtration was ventilated through the ward, so that the stable high concentration of dust was reached. Negative pressure difference was kept in ward related to buffer room. Positive or zero pressure difference was maintained in buffer room related to outdoors, so that there was no disturbance of high concentration outdoors on buffer room. At first this value of negative pressure difference was ignored. Although the negative pressure difference was only less than 2 Pa, the concentration inside buffer room was comparable to that of outdoors, which corresponds to condition 1 in Table 4.4.

Table 2.5 shows the experimental data on the influence of the magnitude of the negative pressure difference on the outward leakage rate of pollution during the opening process of doors. This experiment was carried out in the same laboratory [8]. Enough time was provided for the self-purification of the buffer room, so that it reached the designed ISO 6 air cleanliness level. Then the experiment for the opening and closing of doors within 2 s was performed.

Table 2.2 Illustration of pressure difference incapable of preventing outward leakage of pollution

Situation after opening of door in ward	Door opening in ward, air is fully mixed between buffer room and adjacent room	Through the door in buffer room, 1/10 of air in ward exchanges with air in adjacent room	Through the door in buffer room, no air exchange between ward and adjacent room
Outward leakage of microbes through the buffer room immediately, CFU/year	5×10^6	7×10^4	0

Table 2.3 Relationship between outward leakage of pollution from ward and pressure difference

Pressure difference, Pa	CFU/year
Negative pressure indoors, with door closed	0
Positive pressure indoors, with door closed	
0.001	1.3×10^4
0.01	4×10^4
0.1	13×10^4
1	40×10^4
10	130×10^4
0.001 with door open (area 2 m^2)	2600×10^4

Table 2.4 Small pressure difference may cause large pollution

Condition	Pressure difference between ward and buffer room, Pa		Pressure difference between buffer room and outdoors, Pa		Concentration in buffer room, pc/L	
	Before door opening	During door opening	Before door opening	During door opening	Before door opening	Door opening then closed for 12 s
1	−14	0	−1 to −2	No record	17.1	65,366.8
2	−10	0	+1 to −2	0	12.3	1069.8
3	−10	0	+1 to −2	−1	6.4	1350.6

Table 2.5 Relationship between pressure difference and outward leakage rate during door opening

Pressure difference between ward and buffer room, Pa	Pressure difference between buffer room and external room, Pa	Original particle concentration before opening of door (≥ 0.5 μm), pc/L		Particle concentration in leakage airflow after opening of door (≥ 0.5 μm), pc/L			Note
		Ward (A)	Buffer room (C)	Ward (A)	Buffer room (B)		
					Average	Maximum	
One people exits with door opening and closing for 2 s	−31	+6	75,500	76.2	74,100	1258	2120	Max. value appears at the 2nd minute
	−30	+8	59,440	15.2	57,840	1001	1504	Max. value appears at the 1st minute
	−6	+1	104,240	118	100,706	1854	2832	Max. value appears at the 2nd minute
	0	0	107,580	36.9	109,240	2984	4470	Max. value appears at the 1st minute

(continued)

Table 2.5 (continued)

Pressure difference between ward and buffer room, Pa		Pressure difference between buffer room and external room, Pa	Original particle concentration before opening of door (≥ 0.5 μm), pc/L		Particle concentration in leakage airflow after opening of door (≥ 0.5 μm), pc/L			Note
			Ward (A)	Buffer room (C)	Ward	Buffer room (B)		
						Average	Maximum	
	0	0	96,190	17.1	98,540	3021	5155	Max. value appears at the 1st minute
No people exits or enters with door opening and closing for 2 s	−30	+8	20,517	41.3	20,024	164	271	Max. value appears at the 1st minute
	−6	+4	9358	14.1	10,823	115	196	Max. value appears at the 1st minute
	0	+5	17,073	5.3	15,371	205	267	Max. value appears at the 1st minute
	0	+5	13,498	–	–	150	202	Max. value appears at the 1st minute

In order to avoid the influence of supply air, indoor vortex and opening/closing of doors on the measurement of pressure difference, which should be paid attention to especially for small room, exterior hood should be placed at the original test hole and the vent hole should be set beneath the hood, which is shown in Fig. 2.4.

The following conclusions can be obtained from Table 2.5:

(a) As long as the relative pressure difference between the ward and the buffer room is not positive, there is no leakage of pollution outwards to the buffer room under the condition of door closing, even when the pressure difference is

Fig. 2.4 Set of exterior hood
around the test hold for
measurement of pressure
difference. **a** Test hole for
measurement of pressure
difference. **b** Exterior hood
around the test hole

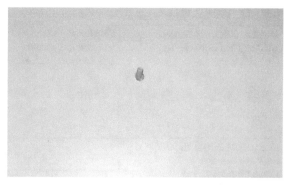

(a) Test hole for measurement of pressure difference

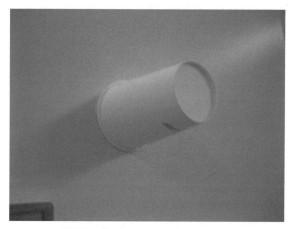

(b) Exterior hood around the test hole

zero. The particle concentration inside the buffer room reached the level comparable with that after self-purification, which was about 0.3% of the concentration in the ward. There was no trend of increase for the particle concentration in the buffer room when the pressure difference is zero. On the contrary, this appears the different trend which was related to the residual influence by door opening, entrance and exit of people.

(b) Under the condition of the same original particle concentration, when the negative pressure difference changed from 0 to −6 Pa, the maximum outward leakage of pollution was converted from 4810 pc/L at 0 Pa to 2832 pc/L at −6 Pa, which was equivalent to the reduction rate of 41%. When the pressure difference changed from 0 to −30 Pa, the reduction rate was 62%. Table 2.6 shows the dimensionless concentration ratio in outward leakage airflow, which was related to the subtraction of the background concentration from the stable concentration in outward leakage airflow. It is shown from Table 2.6 that during the entrance and exit of people, the concentration ratios were

Table 2.6 Relationship between pressure difference and concentration ratio in leakage flow

Pressure difference between ward and buffer room, Pa		Pressure difference between buffer room and exterior space, Pa	Concentration ratio in leakage flow = [concentration in leakage flow (B)-background concentration (C)]/original concentration (A)	
			Based on the average concentration	Based on the maximum concentration
One people exits with door opening and closing for 2 s	−31	+6	0.016	0.027
	−30	+8	0.017	0.025
	−6	+1	0.017	0.026
	0	0	0.027	0.041
	0	0	0.031	0.053
No people exits or enters with door opening and closing for 2 s	−30	+8	0.006	0.011
	−6	+4	0.011	0.019
	0	+5	0.012	0.015
	0	+5	0.011	0.015

Note A, B and C represent the data in Table 2.5

comparable under the pressure difference −6 and −30 Pa. The concentration ratio under the pressure difference −30 Pa was (0.041 + 0.053)/2 = 0.047, while that under 0 Pa was (0.025 + 0.027)/2 = 0.026. The corresponding reduction rate was only 45%, which did not match the variation of pressure difference.

(c) When no people exits or enters with door opening and closing, the difference between the cases 0 and 30 Pa is trivial.

Figure 2.5 vividly shows the relationship between the outward leakage concentration ratio and the pressure difference [9].

If no people enters or exits, the relationship of variation is quite gentle. If people enters or exits, the variation becomes abrupt even when the pressure difference is only less than 6 Pa. This phenomenon is understandable.

The same characteristic appears in the above experiment and the experiment performed by Japanese scholar [1]. In the latter experiment, the operation of addition of 200 g additive was performed in a negative pressure room. After 3 min, it was considered that released particles from operation had been dispersed evenly in the room. Then the particle concentration at the return air opening was measured, until it recovered to the original concentration. Based on the increased concentration value and the flow rate of return air, the released particle number could be obtained. In this experiment, the number of generated particles each time was 2.6×10^8 pc. Then the door was open and closed. Once the door was closed, occupant walked towards the return air opening and measured the particle concentration near the return air opening in the corridor. The method was the same to the above-mentioned one.

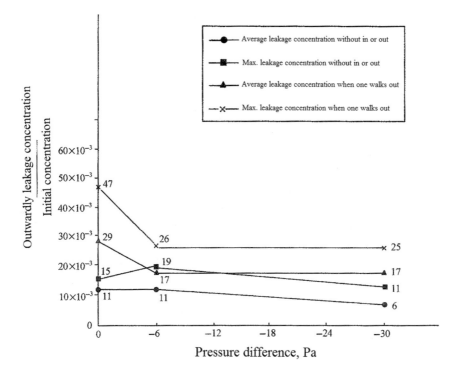

Fig. 2.5 Relationship between outward leakage concentration and pressure difference

The relationship between the pressure difference and the dispersion of pollutant particles is shown in Fig. 2.6.

It is shown in this figure that with the increase of the negative pressure difference value indoors, the number of particles dispersed outwards during opening and closing of doors reduces slightly. The sequence of the number of dispersed particles is that outwardly opening door > inwardly opened door > sliding door.

According to the data provided directly from this literature, when the value of negative pressure difference between indoors and corridor increased from 0 Pa to −30 Pa, the number of particles invading indoors each time during opening and closing of door varied from 4.2×10^6 pc to 1.7×10^6 pc for outwardly opening door, from 1.3×10^6 pc to 1.2×10^6 pc for inwardly opened door, from 0.36×10^6 pc to 0.09×10^6 pc for outwardly opening door, respectively.

It could be estimated from Fig. 2.6 that when the value of negative pressure difference varied from 0 Pa to −6 Pa, the maximum reduction of dispersed particles was about 40%, which is comparable to 41% obtained by our experiment shown in Table 2.5. When the pressure difference changed from 0 Pa to −30 Pa, the reduction of dispersed particles was about 60%, which is comparable to 62% obtained by our experiment shown in Table 2.5. Therefore, it is obvious that no matter whether the pressure difference changes from 0 to −6 Pa, or even −30 Pa, the phenomena of outward leakage of pollution cannot be prevented completely

Fig. 2.6 Number of particles invading indoors during opening and closing of door for negative pressure room

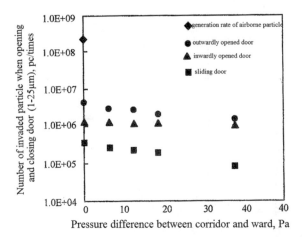

during the opening and closing of doors, and the order of magnitude of the pollution is almost the same.

The test with experimental bacteria also proved this conclusion [8], which will be introduced later in the chapter about buffer room. Colored *B. subtilis* Spores were generated in the ward. The CFUs in the ward were measured when the pressure difference between the ward and the exterior buffer room was −5 and 0 Pa, respectively. The CFU in the buffer room was then measured after people left the ward and entered into the buffer room by opening the door for once time. The temperature differences under two pressure difference values were almost the same. Result showed that the CFU was not inversely proportional to the pressure difference. The influence of the pressure difference on CFU was little, which is shown in Table 2.7.

In order to prevent the leakage of experimental bacteria outwards from the buffer room, negative pressure difference should be maintained between buffer room and outer room. There are no experimental bacteria in outer room. This is different from the before-mentioned test with atmospheric dust.

(2) On the other hand, the pressure difference is not the sole factor for the dispersion of pollutant. Temperature difference also exists for such effect. With daily experience, air flows in and out with the effect of temperature difference,

Table 2.7 Performance of outward leakage of pollution by increasing negative pressure value

Temperature difference between ward and buffer room, °C	Pressure difference between ward and buffer room, Pa	Pressure difference between buffer room and outer room, Pa	CFU in ward (Average from 5 samples), CFU/vessel	CFU in buffer room (Average from 5 samples), CFU/vessel	Concentration ratio
+2.2	0	0	752	185	0.25
+2.1	−5	−5	817.8	179	0.22

which cannot be weakened or offset by pressure difference. This will be explained later in detail. Therefore the pressure difference is not the only measures to prevent the invasion or outward leakage of pollution. Therefore, it is not only the pressure difference which plays a role in the isolation principle inside isolation ward, isolation cleanroom and bio-safety laboratory. As a result, it is indeed a misunderstanding that large negative pressure difference value must be maintained for negative pressure isolation purpose.

2.2 About Airproof Gate

It takes for guaranteed that door must be very air-tight in order to prevent the accidental outward leakage of pollutant airflow from indoors. Here we do not assume that the pressure indoors is positive. Even though the pressure difference indoors is positive and the door is air-tight, exchange of pollution airflow cannot be prevented during dynamic condition, such as opening and closing of doors. Of course, in the static condition, i.e. when the door is closed, exchange of airflow can be guaranteed when sealing with good performance is provided on the gap. However, it should be noted that the door of the ward would be open frequently.

2.2.1 Effect of Entrainment by Door

The influence of door closing and opening on transmission of pollutant between indoors and outdoors is explained as follows.

(1) When the pressure indoors is negative and the outwardly opening door is open, air is pushed by the front face of the door, and the negative pressure formed temporarily in the movement region of door may be much lower than the negative pressure indoors. In this case, indoor air is likely to escape outwardly during the opening of doors.

(2) When the pressure indoors is positive and the inwardly opened door is open, air is also pushed by the front face of the door, and the pressure in the movement region of door becomes negative temporarily. In this case, outdoor air may also be sucked inwardly during the opening of doors.

The above-mentioned phenomenon is termed as the entrainment effect by door. In 1961, Wolfe et al. also paid attention to it [10], which is shown in Fig. 2.7. It was pointed out that the quantity of the sucked air by opening of door for once time in room with positive pressure was about 0.17 m³/s.

Because the negative pressure is formed by the external force from the movement of door opening, there is no air balance between entrance into and exit from

Fig. 2.7 Entrainment effect
by opening of door

the room. Therefore it can be assumed that it does not necessarily sound good for the practice to open door outwardly for negative pressure room, or open door inwardly for positive pressure room.

According to our measurement, the air velocity induced by opening of door without occupant passing through is between 0.15 and 0.3 m/s [4].

Suppose the area for door opening reaches 1.5 m², the quantity of entrained air is between 0.23 and 0.45 m³/s, which is slightly larger than that estimated by Wolfe.

For the airflow with such large velocity, the generated airflow through the whole door opening cannot withstand it by the positive pressure from outdoor air or by the negative pressure from exhaust air.

2.2.2 Dynamic Characteristic of Door [1]

The air tightness of door belongs to the static performance of door, which the entrainment performance of door is mainly related to the dynamic characteristic of door.

The dynamic characteristic of door means the feature of counter current, pressure difference, air velocity at the entrance of door, and the transmission of pollutant between indoors and outdoors during the instantaneous opening and closing of door. The counter current is the airflow from negative pressure room to positive pressure room, or the airflow from low pressure room to the high pressure room.

Based on the observatory laboratory with no temperature difference but with pressure difference shown in Fig. 2.8, Japanese scholars performed investigation on the dynamic characteristic on three kinds of doors shown in Fig. 2.9 (Table 2.8).

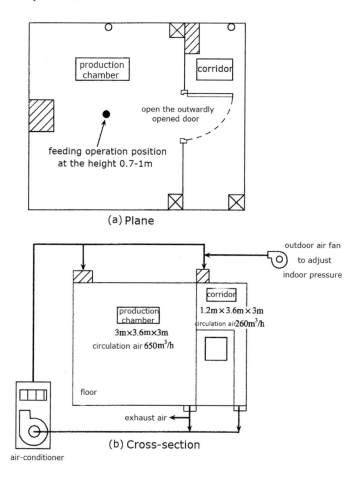

(a) Plane

(b) Cross-section

Fig. 2.8 Observatory laboratory for the dynamic characteristic of door (no temperature difference between indoors and outdoors) ● feeding operation position; ○ differential pressure measurement position; ⊠ return air grille; ⬓ high efficiency air supply outlet

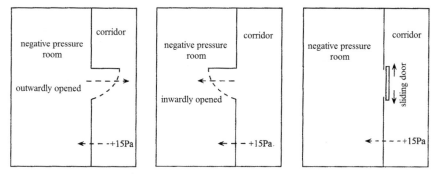

Fig. 2.9 Three kinds of doors

Table 2.8 Situation of visible counter current for three kinds of doors

Door	Opening direction	Stages	ΔP, Pa	Visible counter current
Side hung door	Outwardly	Before door opening During door opening Door opens completely Near door closing fully	15	No Extremely small Slightly weak Strong
	Inwardly	Before door opening During door opening Door opens completely Near door closing fully	15	No Strong No No
	Outwardly	Before door opening During door opening Door opens completely Near door closing fully	0	No Extremely small Slightly weak Slightly strong
Sliding door	Sliding horizontally towards left or right	Various stages	15	Almost invisible

It is shown from the above figure that for the common side hung door, there must exit one phenomenon of strong visible counter current under four stages. This means people are able to see clearly the air flows in the way that it should be prevented. In this case, the isolation performance of air-tight door does not exist. The cost of air-tight door is very high. If interlock function is added, dangerous accident will occur once it is malfunctioned, which has already appeared in the past.

2.2.3   Effect of Entrainment by Occupant

The entrainment by occupant is inseparable from the entrainment by door.

It is found from measurement that the air velocity reaches the maximum during the instantaneous period of door opening when occupant walks in or out. This instantaneous air velocity lasts for about 2 s.

Experiment [11] has shown that when occupant walks along the direction of door opening, the air velocity at the entrance is about 0.14–0.20 m/s; when occupant walks against the direction of door opening, the air velocity becomes 0.08–0.15 m/s. Suppose the dimension of occupant is 1.7 m × 0.4 m, the maximum flow rate by occupant entrainment is 0.14 m^3/s.

This phenomenon is termed as the effect of entrainment by occupant. Wolfe has also discovered this, which is shown in Fig. 2.10 [10].

Fig. 2.10 Effect of
entrainment by entrance or
exit of occupant

2.2.4   Effect of Temperature Difference Between Indoors and Outdoors

It has been introduced in previous chapter on pollution of negative pressure isolation ward that pollution cannot be dispersed outwardly when door is closed.
However, door will be opened and closed frequently. During door opening and
closing, pollution will be exchanged between indoors and outdoors under the
influence of temperature difference.

China Academy of Building Research has performed the specific research on
this aspect. Based on the measurement results by different people, the phenomena
by temperature difference are described under six conditions. There conditions are
shown in Figs. 2.11, 2.12, 2.13, 2.14, 2.15, 2.16, 2.17, 2.18, 2.19, 2.20, 2.21, 2.22,
2.23, 2.24 and 2.25, which include:

$$\Delta t = 0, \quad \Delta P > 0$$
$$\Delta t = 0, \quad \Delta P < 0$$
$$\Delta t > 0, \quad \Delta P > 0$$
$$\Delta t > 0, \quad \Delta P < 0$$
$$\Delta t < 0, \quad \Delta P > 0$$
$$\Delta t < 0, \quad \Delta P < 0$$

The above-mentioned figures are summarized into Table 2.9. Values of Δt and
ΔP at the right workshop are used as baselines [4].

Fig. 2.11 Airflow at the entrance of a room at Guangzhou Ruitai Animal Pharmaceutical Co., Ltd (When door is semi-open, the outwardly airflow at the upper region is much stronger) ($\Delta t = 0$, $\Delta P = +18$ Pa, measured in January 2004)

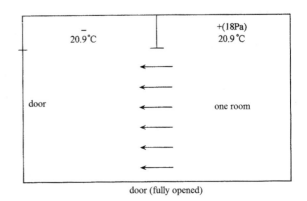

Fig. 2.12 Airflow at the entrance of a buffer room at Liaoning Yikang Biological Co., Ltd ($\Delta t = 0$, $\Delta P = -10$ Pa, measured in May 2004)

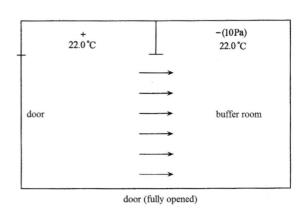

Fig. 2.13 Airflow at the entrance of a room at Guangzhou Ruitai Animal Pharmaceutical Co., Ltd (Air flows inwardly near the floor. Sometimes the flow fluctuates because of the large back flow from floor by large supply air velocity) ($\Delta t = 0$, $\Delta P = -11$ Pa, measured in January 2004)

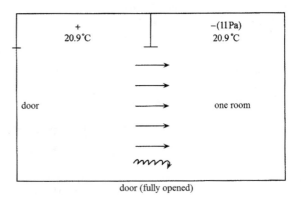

The following conclusions can be obtained from Table 2.9.

(1) When there is no temperature difference between indoors and outdoors, the direction of airflow complies with the direction of the pressure difference. When the pressure indoors is positive, the air flows outwardly, and vice versa.

Fig. 2.14 Airflow at the entrance of a sterile room at Liaoning Yikang Biological Co., Ltd ($\Delta t = 0$, $\Delta P = -12$ Pa, measured in May 2004)

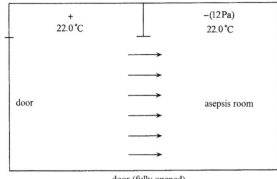

Fig. 2.15 Airflow at the entrance of a room at Guangzhou Ruitai Animal Pharmaceutical Co., Ltd ($\Delta t = 0$, $\Delta P = -50$ Pa, measured in Jan. 2004)

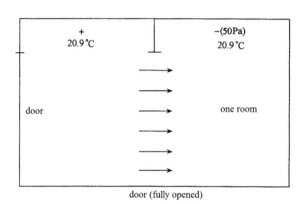

Fig. 2.16 Airflow at the entrance of a refrigeration storage room at Jiangxi Keda Animal Pharmaceutical Co., Ltd (Because the door is close to the return air opening, the airflow velocity at the bottom region is much larger than that at the upper region) ($\Delta t = +0.1$ °C, $\Delta P = +40$ Pa, measured in June 2004)

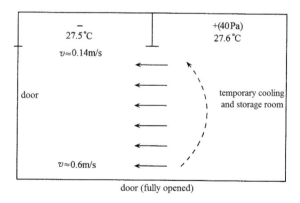

(2) When there is temperature difference between indoors and outdoors, the direction of airflow mainly complies with the direction of the temperature difference. When the temperature indoors is larger than outdoors, or even the temperature difference is as small as 0.1 °C, the air flows outwardly at the upper region and inwardly at the lower region. For the case with small

Fig. 2.17 Airflow at the entrance of a Dressing Room 2 for Woman at Hefei Antewei Animal Pharmaceutical Co., Ltd (There is no obvious inward flow at the bottom of the door under the influence of temperature, and it appears gleamingly) (Δt = +0.4 °C, ΔP =>0 Pa, measured in January 2004)

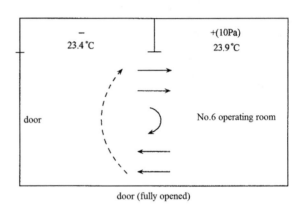

Fig. 2.18 Airflow at the entrance of No. 6 operating room at Chenzhou No. 1 People's Hospital (Δt = +0.5 °C, ΔP = +10 Pa, measured in September 2004, which is explained as sample 8 in Table 2.9)

Fig. 2.19 Airflow at the entrance of a multi-functional cleanroom at Institute of Building Environment and Energy Efficiency, China Academy of Building Research (Δt > 0 °C, the temperature indoors was not recorded, ΔP = +10 Pa, measured in December 2004)

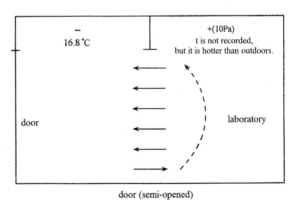

temperature difference, the region of convective airflow is small. When the temperature indoors is less than outdoors, or even the temperature difference is as small as −0.1 °C, the air flows inwardly at the upper region and outwardly at the lower region. For the case with small temperature difference, the region of convective airflow is small.

Fig. 2.20 Airflow at the entrance of a positive pressure room at Zhengzhou Modern Pharmaceutical Co., Ltd ($\Delta t = +1.5$ °C, $\Delta P > 0$ Pa, the pressure indoors was not recorded, measured in September 2004)

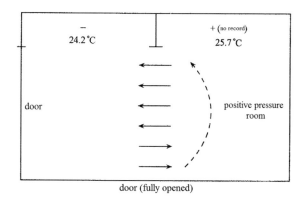

door (fully opened)

Fig. 2.21 Airflow at the entrance of a Dressing Room 1 at Zhenjiang Victor Pharmaceutical Co., Ltd ($\Delta t = -0.1$ °C, $\Delta P = +25$ Pa, measured in May 2004)

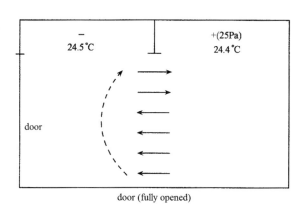

door (fully opened)

Fig. 2.22 Airflow at the entrance of No. 1 operating room at Chenzhou No. 1 People's Hospital ($\Delta t = -0.3$ °C, $\Delta P = +22$ Pa, measured in September 2004)

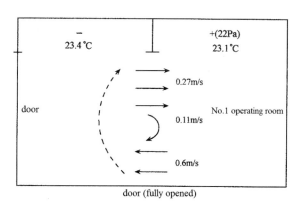

door (fully opened)

(3) Under the influence of various factors, the state of the airflow at the middle of the door may be transitional.

(4) When the directions of airflow dominated with temperature difference is consistent with that with pressure difference, the magnitude of the airflow is strengthened.

Fig. 2.23 Airflow at the entrance of a confecting room at Hefei Antewei Biological&Pharmaceutical Co., Ltd (Δt = −1.3 °C, ΔP = +60 Pa, measured in January 2004)

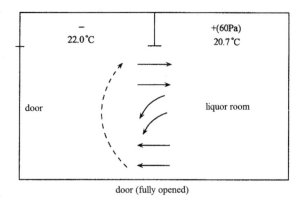

door (fully opened)

Fig. 2.24 Airflow at the entrance of a buffer room at Shangyu Guobang Animal Pharmaceutical Co., Ltd (Δt = −2.9°C, ΔP = +15 Pa, measured in June 2004)

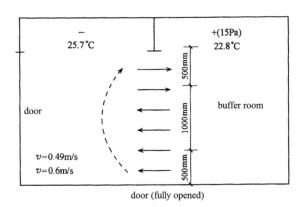

door (fully opened)

Fig. 2.25 Airflow at the entrance of an incubation room at Zhengzhou Modern Pharmaceutical Co., Ltd (Δt = +4.1 °C, ΔP < 0 Pa, the pressure indoors was not recorded, measured in February 2004)

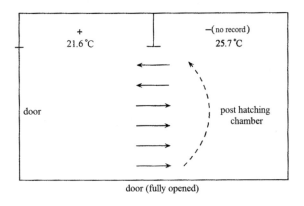

door (fully opened)

(5) When the air supply outlet or return air opening is close to the door, or when the air velocity of supply is so large that it impacts the floor, their influence on convective flow by temperature difference is much larger than the influence by pressure difference.

Table 2.9 Influence of temperature difference on air exchange rate at door during full opening

Group	No.	Δt, °C	Convective airflow	ΔP, Pa	Airflow by pressure difference	Actual airflow condition	Note
I		0		>0			
	1	0	No	+18	Outwardly	Completely outward	
II		0		<0			
	2	0	No	−10	Inwardly	Completely inward	Airflow near the floor fluctuates sometimes, this may be caused by back flow exerted by supply air on floor
	3	0		−11			
	4	0		−12			
	5	0		−50			
III		>0		>0			
	6	+0.1		+40		Completely outward	Airflow at the lower region is influenced by the return air opening outdoors, the sucked air outwardly offsets the inward airflow. So there is no visible inward airflow. The inward airflow at lower region of door appears gleamingly
	7	+0.4		>0 (no record)			The value of Δt is slightly larger than the previous case, so the effect is much stronger than the previous case
	8	+0.5	Outwardly at the upper region, while inwardly at the lower region	+10	Outwardly	Inwardly at upper region, rotating in the central region, outwardly at lower region	Maybe the temperatures were recorded oppositely

(continued)

Table 2.9 (continued)

Group	No.	Δt, °C	Convective airflow	ΔP, Pa	Airflow by pressure difference	Actual airflow condition	Note
	9	>0(no record)		+10		Inwardly near floor, outwardly in other regions	Door was semi-open
	10	+1.5		>0 (no record)		Inwardly at upper region, outwardly in other regions	Because Δt is larger than sample 9, the airflow exchange rate is stronger. So the inwardly airflow rate is slightly larger than sample 9
IV		>0		<0			
	11	+4.1	Outwardly at the upper region, while inwardly at the lower region	<0 (no record)	Inwardly	Outwardly at upper region, inwardly in other regions	Because Δt is larger, there is no transitional flow region
V		<0					
	12	−0.1		+25		Inwardly at upper region, outwardly in other regions	
	13	−0.3		+22		Inwardly at upper region, rotating inwardly in the central region, outwardly in other regions	At the lower region, directions of convective flow and flow by pressure difference are the same. The airflow is strengthened
	14	−1.3	Inwardly at the upper region, while outwardly at the lower region	+60	Outwardly	Inwardly at upper region, horizontally or outwardly inclined in the central region, outwardly at lower region	

(continued)

Table 2.9 (continued)

Group	No.	Δt, °C	Convective airflow	ΔP, Pa	Airflow by pressure difference	Actual airflow condition	Note
	15	−2.9		+15		Inwardly at upper region, outwardly in other regions	Because the magnitude of Δt is larger than sample 13, there is no transitional flow in the central region, and the air velocity in lower region is strengthened to be larger
VI		<0		<0			
	16	No case	Outwardly at the upper region, while inwardly at the lower region	No case	Inwardly	No case	No case

2.2.5 Balance Equation of Air Change Rate with Convective Flow by Temperature Difference

With mechanical ventilation system, the balance equation of air change rate is as follows:

$$Q_1 + Q_2 = Q_3 \tag{2.2}$$

or

$$Q_1 = Q_2 + Q_3$$

where Q_1 is the flow rate of the supply air; Q_2 is the flow rate of the leakage air sucked into or exhausted out of the room through the gap or hole of the gate; Q_3 is the flow rate of the return air or that of the exhaust air.

After the door is open, convective flow occurs under the influence of temperature difference. Indoor cold air flows outwardly through the upper region of the door opening, so there must be the same amount of cold air flowing inwardly through the lower region of the door opening. This can provide the balance between the air change rate and the pressure difference. In this case, we obtain:

$$Q_4 = Q_5 \qquad (2.3)$$

For pure convective flow by temperature difference, the principle of natural ventilation works. The area of intake flow is considered to be the same as that of the outtake flow. In the middle of the door opening, there is a interface where the interior and exterior pressures are the same, which is termed as the neutral plane. It is shown at the position labeled with 0 in Fig. 2.26.

On the intake and outtake flow planes, the pressure difference caused by the density difference can be expressed as:

$$\Delta P = gh\Delta\rho \; (\text{Pa}) \qquad (2.4)$$

where g is the gravitational acceleration, m/s^2; h is the height between centers of intake and outtake planes, m; $h = H/2$, and H is the height of door opening, m; ρ is the density for the intake flow or outtake flow, kg/m^3, which is shown in Table 2.10; $\Delta\rho$ is the density difference of air, kg/m^3.

Because the temperatures of intake and outtake flows are different, the density ρ of hot air is small, so v is large. While the density ρ of cold air is large, so v is small.

Fig. 2.26 Intake and outtake flows through door opening

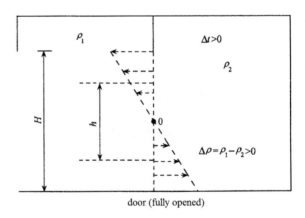

door (fully opened)

Table 2.10 Density of dry air, kg/m^3

t, °C	Atmospheric pressure, mmHg			
	ρ			
	720	740	760	780
−10	1.271	1.307	1.342	1.377
0	1.225	1.259	1.293	1.727
10	1.182	1.215	1.247	1.280
20	1.141	1.173	1.205	1.237
30	1.104	1.134	1.165	1.196
40	1.069	1.098	1.128	1.158

It is shown from Table 2.10 that under the normal atmospheric pressure, when the temperature is within the range 10–30 °C and $\Delta t = 1$ °C, $\Delta \rho = 0.004$ kg/m^3. According to Eq. (2.4), when h = 1 m, we know

$$\Delta P = 9.8 \times 1 \times 0.004 = 0.039 \, \text{Pa}$$

Since the thermal pressure is very small, illusion will be formed that the thermal prepared can be counteracted by the pressure difference easily. However, when door is closed, the intake flow rate by negative pressure difference is exerted on the small area such as door gap. When door is open, the convective flow rate by thermal pressure is exerted on a half of the door opening. When the value of pressure difference is equal to that of the thermal pressure, the former flow rate is much less than the latter. When the former flow rate was exerted on the door opening area, the air velocity formed cannot counteract the air velocity by the latter flow rate. The air velocity on door opening can be expressed as Eq. (3.2) in Chap. 3.

Because the area of the door opening is very large, which is different from the gap, the intake and outtake flow rates (Q_4 and Q_5) can be calculated with Eq. (2.5). The maximum of the parameter φ can be within 0.9 and 0.97, which is different from Sect. 3.1.

$$Q_4 = Q_5 = \varepsilon \varphi F_1 \sqrt{\frac{2\Delta P}{\rho_1}} = \varepsilon \varphi F_2 \sqrt{\frac{2\Delta P}{\rho_2}} \, \left(\text{m}^3/\text{s} \right) \tag{2.5}$$

where F_1 is the area of the outtake flow on the door opening; F_2 is the area of the intake flow on the door opening.

Because the density difference $\Delta \rho$ is very small, it can be simplified for calculation. It is supposed that $\rho_1 = \rho_2$ and $F_1 = F_2$. Equation (2.5) is different from Eq. (2.1) which is used for calculation of the leakage flow rate through door gap. In Eq. (2.1), $\mu = \varepsilon \varphi$, and ε is the contraction coefficient of the flow. But for door opening, $\varepsilon \approx 1$.

2.2.6 Relationship Between Temperature Difference and Pollutant Exchange Rate

According to Eq. (2.4), when the height of the door is 2 m, h = 1 m. The average air velocity and flow rate under different temperature differences are shown in Table 2.11. During the calculation process, $\Delta t = 1$ °C, $\Delta \rho = 0.004$ kg/m^3, the width of door is set 0.9 m, h = 1 m and $\varphi = 0.94$.

It is shown from Table 2.11 that the flow rate of convective flow on door opening by 0.1 °C temperature difference reaches 0.07 m^3/s. It is equivalent to the

Table 2.11 Air velocity and flow rate of convective flow

Δt, °C	0.1	0.2	0.3	0.5	1	1.2	1.5	2.	2.5	3	3.5	4	4.5	5
V, m/s	0.076	0.107	0.13	0.17	0.24	0.26	0.29	0.34	0.38	0.42	0.45	0.48	0.51	0.54
Q_4 (Q_5), m³/s	0.07	0.10	0.12	0.15	0.22	0.24	0.26	0.31	0.34	0.37	0.40	0.43	0.46	0.48

intake flow rate 0.072 m^3/s by negative pressure difference $\Delta P = -15$ Pa as shown in Table 3.1. Therefore, the influence of convective flow by temperature difference cannot be neglected. It is indeed difficult to counteract the influence of temperature difference only by pressure difference when door is open. This is consistent with the aforementioned measured flow direction on door opening.

After door is open, the pollutant exchange rate induced by door opening, movement of occupant and temperature difference is proportional to the flow rate of intake and outtake flow.

The exchange rate of flow by temperature difference is given by Table 2.11.

From previous section, for the area of 1.5 m^2 by door opening (usually the door will not be opened to the vertical position), the maximum flow rate induced by door opening within 2 s is $Q = 0.9$ m^3.

The flow rate induced by movement of occupant within 2 s is $Q = 0.28$ m^3. Therefore, the flow rate without temperature difference is:

$$\sum Q = 1.18\,\mathrm{m}^3$$

The exchange of flow rate within 2 s when there is temperature difference is shown in Table 2.12.

According to the flow rate in Table 2.12, the relationship between the pollutant exchange rates with/without temperature difference can be obtained, which is shown in Fig. 2.27. It is shown that when the temperature difference is larger than 2 °C, the pollutant exchange rate will be greater by 50% from that without temperature difference.

It is shown from Fig. 2.27 that in theory the pollutant exchange rate for $\Delta t = 3$ °C increases by 39% than that for $\Delta t = 0.2$ °C. The pollutant exchange rate for $\Delta t = 5$ °C increases by 55% than that for $\Delta t = 0.2$ °C. But experiment shows that the maximum pollutant concentration (subtracted from the background concentration) for $\Delta t = 5$ °C is 2566.4 pc/L. The maximum pollutant concentration (subtracted from the background concentration) for $\Delta t = 3$ °C is 1622.9 pc/L. The maximum pollutant concentration (subtracted from the background concentration) for $\Delta t = 0.2$ °C is 1378.2 pc/L. Therefore in experiment the pollutant exchange rate for $\Delta t = 3$ °C increases by 18% than that for $\Delta t = 0.2$ °C. The pollutant exchange rate for $\Delta t = 5$ °C increases by 86% than that for $\Delta t = 0.2$ °C [7].

Based on the comparison between experimental data and theoretical data, the difference varies for different temperature differences. However, with the increase of temperature difference, the increasing trend of the pollutant leakage rate is consistent. For example, when the door of the ward opened firstly and then closed, the ratio of average concentration in the buffer room to that in the ward during the first 2 min is shown in Fig. 2.28, which is about 3.5–6.6%.

The relationship between the concentration increase by temperature difference and the time is shown in Fig. 2.29.

Table 2.12 Exchange of flow rate within 2 s with temperature difference

Δt, °C	0	0.1	0.2	0.3	0.5	1	1.2	1.5	2	2.5	3	4	4.5	5
Q, m³/s	1.18	1.32	1.38	1.42	1.48	1.62	1.66	1.70	1.8	1.86	1.92	2.04	2.1	2.14

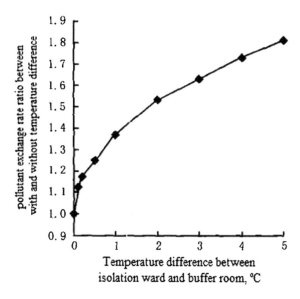

Fig. 2.27 Pollutant leakage rate by convective flow with temperature difference before and after door opening

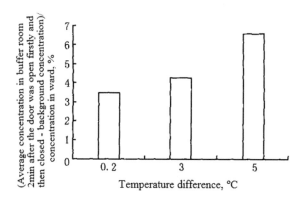

Fig. 2.28 Average concentration in buffer room 2 min after the door was open firstly and then closed

The phenomenon of aforementioned convective flow by temperature difference and its effect on pollutant exchange rate on door opening are rare in foreign standards and literatures. In *"Guidelines for Preventing the Transmission of Mycobacterium tuberculosis in Healthcare Facilities"* issued by the Centers for Disease Control and Prevention (CDC) in U.S.A. in 1994, it was only mentioned that "indoor air distribution is affected by the temperature difference of air", but the influence of the temperature difference on the pollutant exchange was not noticed. In ASHRAE manual, this has been paid attention to. But it was only mentioned that when the door was open, "because convection occurs with temperature difference between two areas, air exchange by natural ventilation appears". There is no deep

Fig. 2.29 Variation of
concentration in buffer room
3 min after the door was open
firstly and then closed

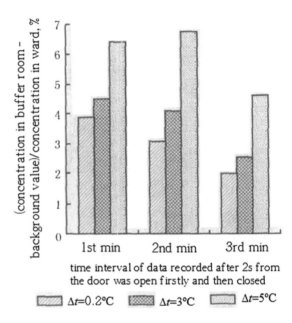

and quantitative analysis. The observatory results by foreign scholars have also
been provided, which are shown in Figs. 2.30 and 2.31. But there is no further
investigation.

In Fig. 2.30, the temperatures in the ward and in the corridor were 21 and 22 °C,
respectively. Although air flow kept balanced at 1 m above the floor near the door
opening, this is obviously the situation when door is open. But the velocity of air
flow near the floor induced by temperature difference is near 0.3 m/s (equivalent to

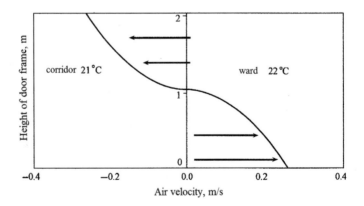

Fig. 2.30 Convection on door opening when $\Delta t = 1$ °C

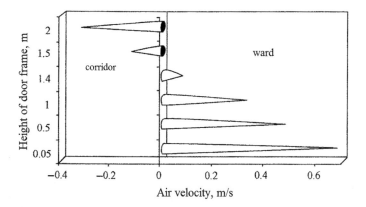

Fig. 2.31 Convection on door opening when $\Delta t = 3.6\ ^\circ C$

600 m³/h), which is in the opposite direction of the air flow in the upper region of door opening. Someone believed that when the air change rate could not reach 10 h⁻¹ or 1200 m³/h (when the original 600 m³/h of the outtake flow in the upper region of door opening, or that of the intake flow in the lower region of door opening, needs to be counteracted, there should be the corresponding intake flow with flow rate 600 m³/h in the lower region, or the corresponding outtake flow with flow rate 600 m³/h in the upper region, respectively. Therefore it is 1200 m³/h), pollutant will escape to the corridor. But this researcher did not pay attention to the impossibility for the formation of such a large exhaust flow in the ward. Even in the normal condition, the pressure difference is unable to counteract the convection by temperature difference. This conclusion has not been proposed. On the contrary, until now it is believed that the pressure difference is able to counteract the convection by temperature difference (refer to the section about pressure difference).

Figure 2.31 shows the flow direction on the cross section of the door with width 900 mm, when the temperature difference between two rooms reached 3.6 °C. It is shown that the air change rate at the height 1.45 m above the floor was almost 0, where can be considered to be the neutral plane. The air velocity in the lower region of the door opening from the corridor to the ward reached 0.7 m/s. The air velocity in the upper region of the door opening reached 0.3 m/s with the opposite direction. It was also pointed out that with the increase of exhaust air from the ward, the height of the neutral plane would move upwards.

Based on the above-mentioned description, experiment and calculation on several phenomena of door, no matter how airproof it is for the door of the isolation ward, the intake and outtake of air flow under dynamic condition cannot be prevented. Therefore, the airproof door cannot play the role of effective isolation.

Therefore, it is again a misunderstanding that airproof door must be installed for negative pressure isolation ward (In extreme case, the door interlock is used, or even the airproof door used in submarine is applied).

2.3   About Full Fresh Air

2.3.1   Outline

After SARS epidemic, the design personnel are inclined to adopt the full fresh air scheme during the design of the negative pressure isolation ward. It is believed that the circulation air cannot prevent the air pollution.

But when the related regulations abroad are referred, except the ASHRAE manual "*Health Care Facilities*" issued in 1991 and the Russian standard, full fresh air scheme is not necessary in other guidelines and the above-mentioned literatures revised after 2003, which is shown in Table 2.13. In the ASHRAE manual, apparatus with circulation air is not allowed. In the Russian standard, all the air should be exhausted outdoors.

In the guidelines issued by CDC in U.S.A., there are two situations when full fresh air system is not necessary and circulation air is permitted:

(1) Multi-room system. When air from these rooms should be recirculated to the HVAC system, HEPA filters should be installed at the individual exhaust (or return) air pipelines, instead of the total exhaust pipeline.
(2) Single room. When three kinds of installation types for HEPA filter are allowed: (a) HEPA filter is installed at the return air pipeline of the room, so that return air is filtered and then delivered into the room; (b) when HEPA filter is installed on wall or above the ceiling, indoor self-circulation air system is formed, so that the filtered air can be used again; (c) HEPA filter is installed in the filtration unit, but it is not specified by CDC that this kind of the filtration unit should be installed above the ceiling or in the room.

2.3.2   HEPA Filter and Virus Particles

The main concern about the scheme to use full fresh air only and no circulation air in negative pressure isolation ward is that there may be pathogenic microbes with enough concentration in recirculation air. Since the size of virus is very small, it is believed that they cannot be removed by HEPA filter. It is also considered that there are basically no pathogenic bacteria in fresh air.

The belief that virus cannot be removed by HEPA filter means that the virus size is too small. Indeed, the size of virus itself is too small, which is about 0.01–0.1 μm. But actually the virus is attached to its nutritious matters. They are released into the air by various forces such as coughing. This means that there are carriers for the existing virus (also including bacteria) in air. The diameter of these microbes with carriers was termed as the equivalent diameter [11]. For example, the size of the bacteriophage in the processing of the monosodium glutamate factory is about 2–5 μm. Selleris and Herniman investigated the foot-and-mouth disease virus (FMDV) in the natural world. Although the size itself is only 25–30 nm, after

Table 2.13 Related standards on flow rates of dilution air and fresh air

Standard	Specified flow rate for dilution air and fresh air
"Guidelines for Preventing the Transmission of Mycobacterium tuberculosis in Healthcare Facilities" issued by CDC in U.S.A., ASHRAE 170 "Ventilation of Health Care Facilities"	In new-built or renovated isolation ward for prevention of airborne transmission, the air change rate >12 h^{-1} and the fresh air volume >2 h^{-1}
ASHRAE manual (2003) "Health Care Facilities" [14]	In isolation ward, the flow rate of dilution air >6 h^{-1} (based on requirement for odor and thermal comfort)
UK "Guidance on the prevention and control of transmission of multiple drug-resistant Tuberculosis" [14]	In new-built or renovated isolation ward for prevention of airborne transmission, the flow rate of dilution air $\geq$ 12 h^{-1} and the fresh air volume $\geq$ 2 h^{-1}
CDC in U.S.A. "Guidelines for environmental infection control in health care facilities" [14]	In newly-built ward, the air supply volume $\geq$ 12 h^{-1}. In existing ward, the air supply volume $\geq$ 6 h^{-1}
AIA in U.S.A. "Guidelines for design and construction of hospital and health care facilities" [14]	In isolation ward for prevention of airborne transmission, consulting rooms for emergency or radiotherapy, the flow rate of dilution air $\geq$ 12 h^{-1} and the fresh air volume $\geq$ 2 h^{-1}
DHHS in U.S.A. "Guidelines for construction and equipment of hospital and medical facilities" [14]	In isolation ward for prevention of airborne transmission, the air supply volume $\geq$ 12 h^{-1} and the fresh air volume $\geq$ 2 h^{-1}. In the bathroom, laundry, waste disposal room, disinfection room, anteroom of isolation ward, the exhaust air volume $\geq$ 10 h^{-1}
Australia "Guidelines for the classification and design of isolation rooms in health care facilities" [14]	In negative pressure isolation ward, the flow rate of dilution air should be the larger value between 12 h^{-1} and 522 m^3/h
USA "Guidelines on the design and operation of HVAC systems in disease isolation areas" [14]	In newly-built isolation ward, disposal room and mortuary, the flow rate of dilution air $\geq$ 12 h^{-1}. In the bathroom, the exhaust air volume $\geq$ 10 h^{-1}. In the consulting room for infectious patient, the flow rate of dilution air $\geq$ 15 h^{-1} and the fresh air volume $\geq$ 2 h^{-1}
"Guideline for Design and Operation of Hospital HVAC Systems" (HEAS-02-2004) established by Healthcare Engineering Association of Japan	In isolation ward, the total air supply volume should be 12 h^{-1}, and the fresh air volume $\geq$ 2 h^{-1}
DIN 1946-4 standard "Ventilation and air conditioning - Part 4: Ventilation in buildings and rooms of health care" by Deutsches Institut Fur Normung E.V.	In isolation ward, reception room and ICU, the fresh air rate per person is 40 m^3/h. In the ward containing isolated patients with infectious potential, the fresh air rate per person >100 m^3/h (it is known that it should be the circulation air)
Russian standard GOST R 52539-2006 "Air cleanliness in hospitals. general requirements"	In infectious isolation ward (where local air cleaning equipment can be installed), the air change rate should be 12–20 h^{-1}, and total air is exhausted outdoors

Table 2.14 Comparison of penetration for filters

Filter	Flow rate, m³/h	Pressure difference, Pa	Penetration, %			
			Bacteriophage, 0.1 μm	Virus, 0.3 μm	FMDV, 0.01– 0.012 μm	DOP, 0.3 μm
HEPA A	42.5	264	0.0039			0.011
HEPA B	42.5	175	0.00085			0.02
HEPA C	42.5	135	0.00085			0.006
HEPA D	–	–	0.003	0.0036	0.001	0.01

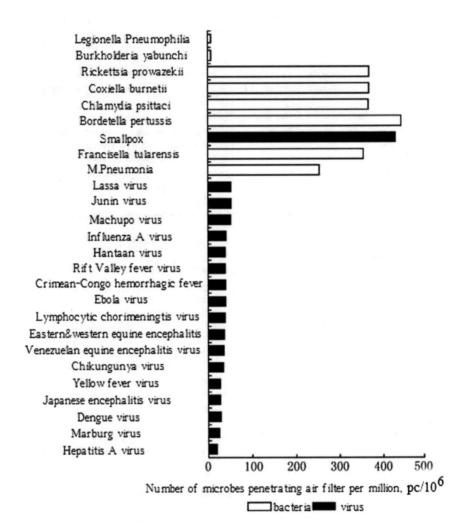

Fig. 2.32 Penetration of microbes through air filters

sampled with cascade liquid impactor, 65–71% of the airborne particles containing virus have sizes larger than 6 μm, and 19–21% have sized between 3 and 6 μm, and only 10–11% have sizes smaller than 3 μm. Therefore, is the number of particles containing virus with size 0.01–0.1 μm (or the virus particles with size 0.01–0.1 μm) too less [12]?

Experiment has shown that the efficiency of HEPA filter for 0.1 μm particles is much larger than that for 0.3 μm, which is shown in Table 2.14 [10]. This also means that the equivalent diameter of virus is much larger than 0.3 μm.

Based on the above-mentioned introduction, it does not mean that the filtration efficiency for bacteria is large, while that for virus is small. Figure 2.11 shows that the filtration efficiency for may virus is much larger than that for bacteria [13] (Fig. 2.32).

In total, the belief to use full fresh air only and no circulation air in negative pressure isolation ward is the third misunderstanding.

References

1. S. Honda, Y. Kita, K. Isono, K. Kashiwase, Y Morikawa, Dynamic characteristics of the door opening and closing operation and transfer of airborne particles in a cleanroom at solid tablet manufacturing factories. Trans. Soc. Heating, Air-Conditioning Sanit. Eng. Japan, **95**, 63–71 (2004)
2. ASHRAE, *ASHRAE Handbook—HVAC Applications* (2003)
3. Z. Xu, *Fundamentals of Air Cleaning Technology*, 3rd edn. (Science Press, Beijing, 2003), pp. 301–304
4. Z. Xu, Y. Zhang, Q. Wang, H. Liu, F. Wen, X. Feng, Isolation principle of isolation wards. *J. HV&AC* **36**(1), 1–7, 34 (2006)
5. F. Chan, V. Cheung, Y. Li, A. Wong, R. Yau, L. Yang, Air distribution design in a SARS ward with multiple beds. Build. Energy Environ. **23**(1), 21–33 (2004)
6. O.C. Ruge, H. Banrud, O. Bjordal, Ultraviolet technology and intelligent pressure control solutions jointly provide true isolation rooms. Klean ASA (2000)
7. Z. Xu, Y. Zhang, Q. Wang, H. Liu, F. Wen, X. Feng, Y. Zhang, L. Zhao, R. Wang, W. Niu, D. Yao, X. Yu, X. Yi, Y. Ou, W. Lu, Study on isolation effects of isolation wards (3). J. HV&AC **36**(5), 1–4 (2006)
8. Z. Xu, Y. Zhang, Q. Wang, F. Wen, H. Liu, L. Zhao, X. Feng, Y. Zhang, R. Wang, W. Niu, Y. Di, X. Yu, X. Yi, Y. Ou, W. Lu, Study on isolation effects of isolation wards (1). J. HV&AC **36**(3), 1–9 (2006)
9. Z. Xu, Y. Zhang, Q. Wang, F. Wen, H. Liu, L. Zhao, X. Feng, Y. Zhang, R. Wang, W. Niu, Y. Di, D. Yao, X. Yu, X. Yi, Y. Ou, W. Lu, Study on isolation effects of isolation wards (2). J. HV&AC **36**(4), 1–4 (2006)
10. H.W. Wolf, M.H. Havris, L.B. Hall, Open operating room doors and staphylococcus aureus. *Hospitals (Lond).* **35**, 57 (1961)
11. Z. Xu, *Fundamentals of Air Cleaning Technology and Its Application in Cleanrooms* (Springer Press, New York, 2014), p. 428
12. X. Yu, Air cleaning : the major measure for removing microbial aerosol particles. J. HV&AC, **41**(2), 32–37 (2011)
13. D. Verreault, S. Moineau, C. Duchaine, Methods for sampling of airborne viruses. Microbiol. Mol. Biol. Rev. **72**(3), 413–444 (2008)
14. J. Shen, Multi-application isolation ward and its air conditioning technique without condensed water. Build. Energy Environ. **24**(3), 22–26 (2005)

Chapter 3
Principle and Technology of Dynamic Isolation

Based on the reflection on these three misunderstandings for the design of negative pressure isolation ward, as well as a series of experimental studies, both the concept and the effective technology for effective isolation of airborne transmission were formed under the dynamic condition during the common operation of negative pressure isolation ward. The dynamic isolation concept proposed by author in 2002 [1] can be established completely [2]. The dynamic isolation is relative to the static isolation which means the application of the barrier (such as closing the air-proof door) and the static pressure difference to prevent leakage through gap. This concept has been adopted by Beijing local standard DB11/663-2009 "*Essential construction requirements of negative pressure isolation wards*". It has also been applied in the design and the construction of negative pressure isolation ward in different places.

3.1 Proper Pressure Difference for Isolation

According to the principle of air cleaning technology, the purpose of isolation is to prevent infection and cross infection. It is especially applied to prevent the transmission of infectious pathogen between indoors and outdoors through air movement. It is an effective measure to realize the purpose of infection control.

For example, the following aspects are considered as the main reasons for the prolonged period of infection by pulmonary tuberculosis, including the delayed treatment on the tuberculosis patients, shortage of isolation period, and deficiency of ventilation in isolation ward.

Except for isolation with barrier (physical isolation) such as isolation room, the terminology of isolation mentioned here mainly means the isolation with pressure difference. This has been illustrated in the previous chapter. It will be unsuccessful to rely on the pressure difference to realized dynamic isolation. The pressure difference is aimed to maintain static isolation. Therefore, in the concept of dynamic isolation, isolation with proper pressure difference is needed.

© Springer Nature Singapore Pte Ltd. 2017
Z. Xu and B. Zhou, *Dynamic Isolation Technologies in Negative Pressure Isolation Wards*, DOI 10.1007/978-981-10-2923-3_3

3.1.1 Physical Significance of Pressure Difference

When all the doors and windows indoors are closed, the pressure difference is the resistance of airflow through the gap of the closed door or window, which is from the high pressure towards low pressure.

From Fig. 3.1, when the pressures at both sides of the gap are assumed P_1 and P_2, the pressure difference can be expressed as:

$$\Delta P = P_1 - P_2 = (\xi_1 + \xi_2)\frac{v^2 \rho}{2} + h_w (\text{Pa}) \qquad (3.1)$$

where

$(\xi_1 + \xi_2)\frac{v^2 \rho}{2}$ is the local resistance at the gap where air flows through;

h_w is the frictional resistance. Since the depth of the gaps on door, window and panel are in the magnitude of 10^{-2} m, h_w can be ignored completely;

ξ_1 is the local resistance at the sudden contraction position. Since the cross sectional area of the gap is extremely small, $\xi_1 \approx 0.5$;

$\xi2$ is the local resistance at the sudden expansion position. Since the cross sectional area of the gap is extremely small, $\xi1 \approx 1$;

Fig. 3.1 Schematic diagram of gap

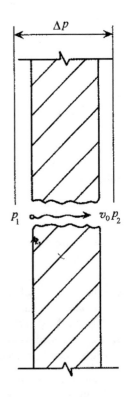

ρ is the air density, which is about 1.2 kg/m^3

From Eq. (3.1), we can obtain

$$v = \frac{1}{\sqrt{\xi_1 + \xi_2}} \sqrt{\frac{2\Delta P}{\rho}} = \varphi \sqrt{\frac{2\Delta P}{\rho}} \, (\text{m/s}) \tag{3.2}$$

where φ is the velocity coefficient, $\varphi = \frac{1}{\sqrt{\xi_1 + \xi_2}} = 0.82$.

Because the resistances of the gaps are different, the value of φ could be large or small. The air velocity of leakage air at different parts of the gap will be different.

Because the geometry of the gap is relatively complex, its resistance will be increased. The value of the velocity coefficient φ will be reduced. With the given ΔP, the air velocity through the gap will be decreased, which will be introduced later.

3.1.2 Determination of Pressure Difference

How to determine the pressure difference for the isolation ward under the condition of door and window closing? It has already been pointed out that it should depend on the enough flow rate of outdoor air to be sucked in through the gap of the door, so that the leakage pollutant airflow through the door gap can be prevented [3]. This belongs to the static isolation.

According to the field test by author, the maximum air velocity induced by occupant movement is 0.34 m/s when the walking velocity is 1 m/s.

The air velocity indoors created by air supply from air conditioner is usually not larger than 0.3 m/s.

The air velocity in normal room with natural ventilation is not larger than 0.2 m/s, which means that the air velocity induced through the gap will not be larger than 0.5 m/s.

From Eq. (3.2), when the air velocity through the gap is 0.5 m/s, the theoretical pressure difference can be calculated with:

$$\Delta P = \frac{\rho v^2}{2\varphi^2} = \frac{1.2 \times 0.5^2}{2 \times 0.82^2} = 0.22 \, \text{Pa}$$

This means that in theory when the door is closed and the pressure difference reaches 0.22 Pa, the requirement for common leakage prevention can be satisfied. It has been pointed out early that when $\Delta P = 1$ Pa the air velocity through the gap in theory could reach 1.06 m/s [4], which could counteract the leakage completely. After the epidemic of SARS, it has been pointed out by "*Guidelines for Preventing the Transmission of Mycobacterium tuberculosis in HealthCare Facilities*" published in 1994 by CDC that for maintenance of the negative pressure and prevention of air flow into the ward, the minimum pressure difference is extremely small (0.001 in H$_2$O). It was considered that with the pressure difference 0.001 in H$_2$O, the leakage flow rate could reach 50 ft^3/min (85 m^3/h). In this case, the minimum air velocity of leakage air sucked inwardly is 100 ft/min, which corresponds to 0.51 m/s.

The static pressure 0.001 in H$_2$O corresponds to 0.25 Pa. With Eq. (3.2), the corresponding ideal air velocity through the gap is 0.53 m/s.

Table 3.1 shows the requirements of pressure difference in isolation ward from standards abroad. In U.S.A., the pressure difference increases from the initial value 0.25 Pa to the current value 2.5 Pa. The reason why the pressure difference should be increased by ten time is not explained.

Are the real air velocity through the gap and the pressure difference consistent with the above-mentioned condition? The necessary minimum pressure difference will be analyzed further.

Table 3.1 Requirements of pressure difference in isolation ward from standards abroad

Standard or guideline	Control object	Negative pressure difference between the ward and the corridor (buffer room), Pa
CDC guideline in U.S.A. (1994)	Mycobacterium tuberculosis	0.25
ASHRAE handbook (*Health care facilities*) in U.S.A. (2003)	Not specified	0.25
UK "*guidance on the prevention and control of transmission of multiple drug-resistant tuberculosis*" [27]	Mycobacterium tuberculosis	0.25
CDC in U.S.A. "*guidelines for environmental infection control in health care facilities*" [27]	Not specified	2.5
DHHS in U.S.A. "*guidelines for construction and equipment of hospital and medical facilities*" [27]	Not specified	2.5
AIA in U.S.A. "*guidelines for design and construction of hospital and health care facilities*" [27]	Not specified	2.5
Australia "*guidelines for the classification and design of isolation rooms in health care facilities*" [27]	Aerosol	15
ASHRAE 170 "*ventilation of health care facilities*" (2013)	Isolation ward with airborne infection	2.5
Russian standard GOST R 52539-2006 "*air cleanliness in hospitals. general requirements*"	Isolation ward with airborne infection	10–15

The case with $\varphi = 0.82$ belongs to the ideal condition for the gap. In fact, the resistance on the gap is much larger. Table 3.2 shows the air velocity through the door gap in real situation. Of course, it is difficult to perform measurement. Therefore, there is error. But from these data, the credibility of the theoretical equation can be found.

In case 7 from Table 3.2, the actual measured pressure difference is 0. The air velocity through the door gap should be zero, but the measured value is 0.18 m/s. This means there is measurement error for the pressure difference 0 Pa, which could not be used to calculate the actual velocity coefficient.

In the previous 6 cases from Table 3.2, the average value is $\overline{\varphi} = 0.29$. When it is used as the velocity coefficient for case 7, we obtain:

$$0.18 = 0.29\sqrt{\frac{2\Delta P}{1.2}}$$

The actual pressure difference for case 7 is $\Delta P = 0.23$ Pa.

The pressure difference value 0.23 Pa is much smaller than the half of the resolution of the current liquid column manometer, i.e., 1 Pa. It is immeasurable, not alone to say the needle manometer. Table 3.2 shows the measured pressure difference was 0, which is natural. This measured result 0 Pa does not represent the actual pressure difference.

From Table 3.2, the actual value φ is between 0.2 and 0.5. Suppose it was 0.5 (when the air-tightness level of doors is less than that of 0.2), when $v = 0.5$ m/s, $\Delta P = 2.6$ Pa. This means that both the theoretical calculation result and the pressure difference 0.23–0.25 Pa provided by CDC from U.S.A. are not practical. It is not only unsafe, but also difficult to measure and control automatically. In fact, the more air-tightness the geometry is, the larger the resistance of the gap is. In this case, the necessary ΔP is much larger. When the air velocity through the door gap is required to be larger than 0.5 m/s, the minimum pressure difference in theory should be larger than 3 Pa.

It is shown that when the door and the window are closed, the pressure difference to prevent the leakage through the gap could be as small as 3 Pa. The opinion is unnecessary that the larger the pressure difference is, the better it is, which will be explained with the experimental data later. But when the pressure difference is as small as 1 Pa, the requirement cannot be satisfied.

Therefore, the following two concepts are provided:

(1) The pressure difference of the room needed is not large. It is feasible to adopt the common value 5 Pa.

The reason why the pressure difference +5 Pa is adopted in common cleanroom will be introduced. One is that it can meet the requirement. The other is that 5 Pa corresponds to 0.5 mm H_2O. It is the half of the smallest scale of the manometer in the metric system, which means the resolution is 0.5 mm. Therefore, in imperial unit system the half of the smallest scale is adopted as the minimum pressure difference, which is 0.05 in H_2O or 1.27 mm H_2O or 12.5 Pa.

Table 3.2 Measured air velocities through door gaps

No.	Location	Door gap, mm		Pressure difference, Pa	Air velocity through gap, m/s		Theoretical air velocity though gap for $\varphi = 0.82$	Actual velocity coefficient φ	Year
		Depth	Length		Measured value	Average			
1	No. 1 operating room at Binzhou people's hospital	20	1500	+22	2.80 3.0 3.0 3.40 3.90	3.22	4.97	0.53	2004
2	No. 1 operating room at Zhengzhou people's hospital	20	1500	+10	1.00 0.65 0.45 0.65 0.75	0.7	3.34	0.172	2004
3	No. 1 Asepsis room at inner Mongolia biological & pharmaceutical Co., Ltd	8	960	+7	0.55 0.60 2.50	1.22	2.8	0.36	2004
4	No. 3 Asepsis room at inner Mongolia biological & pharmaceutical Co., Ltd.	8	960	+1	0.3 0.28 0.2	0.26	1.06	0.25	2004
5	Bacteria Room at inner Mongolia biological & pharmaceutical Co., Ltd.	5	960	−7	0.85 0.80 0.70	0.78	2.8	0.23	2004
6	Jiangxi Keda animal pharmaceutical Co., Ltd.	/	/	+4	0.48 0.52 0.56 0.49 0.51 0.48	0.51	2.11	0.2	2004
	Average velocity coefficient							(0.29)	
7	Chemical analysis lab at inner Mongolia biological & pharmaceutical Co., Ltd.	5	960	0	0.17 0.20 0.18	0.18	0	(0.29)	2004

For the purpose of automatic control, in order to prevent too much instantaneous fluctuation of the pressure difference, it is impossible to maintain 5 Pa, so it is possible to use 10 Pa. When the relative pressure difference is too much, such that larger than 30 or 50 Pa, the people or the little creature will feel uncomfortable. Therefore, in the *"Requirements for ultra-clean ventilation (UCV), Systems for operating departments"* published by National Health Service in U.K. and National Institute for Health Research in U.K., as well as *"Architectural technical code for hospital clean operating department"* (GB 50333-2002) in China, the pressure difference is specified not to exceed the limit of 30 Pa. In the later revision of GB 50333 issued in 2013, the pressure difference reduces from original 30 to 20 Pa based on the requirement of ISO 14644.

(2)   Trouble maybe occur when the envelope of the room is extremely air-proof.

This has also been discovered by literature [5]. It has been pointed out in this paper that when the automatic air valve at the exhaust air pipeline changes the position slightly because of the control error, although the variation of air volume induced is very small, the influence on the fluctuation of indoor pressure will be very large.

During the adjusting process on negative pressure isolation room, when the variation of air volume for the room with the sealing strip near the edges of the door is only several m^3/h, which corresponds to one thousandth of the exhaust air volume from the room, the change of the pressure difference could reach 1 Pa. For example, when the air volume in a lab (23.9 m^2 × 3 m) changed by 5–10 m^3/h, the pressure difference varied by 1 Pa [6]. In order to stabilize the pressure difference, the sealing strip was forced to be removed. It is common that the variation of the air volume reaches several m^3/h.

3.2   Buffer Room for Isolation

From the aforementioned analysis, no matter whether the door is air-proof, with the influence of door opening, occupant movement and the temperature difference, pollutant will release outwardly with the same magnitude during the opening process of door. However, when door is closed and the pressure difference is as small as 5 Pa, the velocity of the entrained air at the gap could reach more than 2 m/s, which could prevent the outward leakage of pollutant. The leakage rate reduced by 40 and 60% when the pressure difference is −6 and −30 Pa, respectively. Therefore, the concept of buffer room to prevent the outward leakage of pollutant efficiently will be proposed.

3.2.1 Mode of Buffer Room

1. Basic mode

Buffer room is the air lock room where clean air is supplied through HEPA filter.

Figure 3.2 shows the schematic diagram of an air lock room. It acts as an auxiliary part of the cleanroom. It was firstly proposed by "*Contamination Control of Aerospace Facilities*" (TO 00-25-203) issued by U.S. Air Force in 1961. Air lock room is a small room near the entrance of the cleanroom. For several doors in the air lock room, only one door could be open at the same time. It is aimed to prevent the contaminated air in the outside area from flowing into the cleanroom, so that the "sealing" function works.

Of course, the air lock room could also be used to prevent the contaminated air inside the room from flowing into the environment.

In "*Good Manufacturing Practice*" (GMP) by WHO, "Airlock is an enclosed space with two or more doors, which is interposed between two or more rooms, e.g. of differing classes of cleanliness, for the purpose of controlling the airflow between those rooms when they need to be entered. An airlock is designed for use either by people or for goods and/or equipment".

It is shown that air lock room is only a room with interlock doors. It is the same as the delivery window. When its volume is not large, the maximum quantity of polluted air for the other side is equivalent to one fourth of its volume. But this kind of polluted air is different from that enters into the buffer room, which has not been diluted by clean air.

Air lock room can only control airflow, but cannot dilute airflow.

Gradient pressure difference is established between two adjacent connected areas, which reduces the pressure value from the pollutant prevention side to the polluted side. In this way, pollution through the gap between two regions (rooms) by induction of some factor can be prevented, which moved from the polluted side to the pollution prevention side.

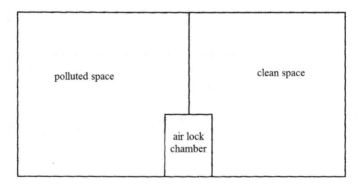

Fig. 3.2 Schematic diagram of an air lock room

In general, the area with high pressure for pollution prevention and the area with low pressure for isolation should be place at the end or at the center of the plane, which is shown in Fig. 3.3. The pressure difference of the isolation ward relative to the atmosphere could be positive or negative. In this book, it is mainly aimed for negative pressure.

Buffer room is placed outside of the isolation ward. Positive pressure is maintained in the buffer room relative the isolation ward, while negative pressure or zero pressure is kept in buffer room relative to the outside of the buffer room. This kind is called Three-Room-One-Buffer, or Two-Area-One-Buffer, which is shown in Fig. 3.4. Three rooms mean the isolation ward, the buffer room and the corridor. Two areas mean the polluted isolation ward and the corridor with potential pollution.

Buffer room is placed outside of the isolation ward. Inner corridor is set outside of the buffer room. The second buffer room is set outside of the inner corridor. Preparatory area for medical personnel is set outside of the buffer room. Positive pressure or zero pressure is maintained in the preparatory area. Negative pressures are kept inwardly. The magnitude of negative pressure increases gradually. This type is called Five-Room-Two-Buffer, or Three-Area-Two-Buffer, which is shown in Fig. 3.5. Five rooms mean the isolation ward, the buffer room 1, the inner

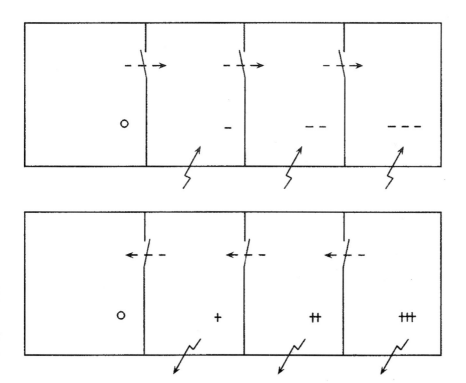

Fig. 3.3 Gradient pressure difference on the plane

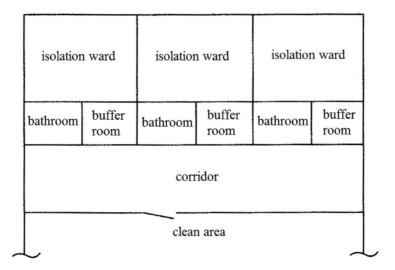

Fig. 3.4 Schematic diagram of three-room-one-buffer

Fig. 3.5 Schematic diagram of five-room-two-buffer

corridor, the buffer room 2, and the clean area. Three areas mean the polluted area, the area with potential pollution, and the clean area.

The isolation ward belongs to the polluted area. The inner corridor belongs to the semi-polluted area. The preparatory area belongs to the clean area.

2. Analysis on pollutant flux

After theoretical analysis on the effect of the buffer room on the negative pressure isolation ward, quantitative assessment of the pollution flux and the effect of the buffer room was obtained. Novel founding was provided for the effect of the buffer room [7–9].

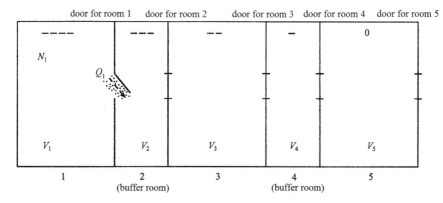

Fig. 3.6 Phase diagram calculation of five-room-two-buffer

Figure 3.6 shows the phase diagram calculation is provided for analysis of pollutant flux. In the figure, No. 1–5 is the serial number of the room. V is the room volume. N_1 is the pollutant concentration in Room 1 or at the gate of Room 1, pc/m^3. Q_1 is the flow rate from Room 1 to Room 2 because of the pressure difference which is not counteracted after door is open, m^3.

Now the analysis on the pollutant flux is performed as follows.

(1) At the moment of door opening for Room 1, the pollutant flux at the gate is $N_1 Q_1$.

(2) The volume of the buffer room is very small (it is usually not larger than 5–6 m^3) and the air change rate is very large (it is usually about dozens h^{-1}). With the effect of dispersion, during the 2–3 s for opening and closing of door for Room 1, three conditions of pollutant distribution which enters into Room 2 can be assumed, which is shown in Fig. 3.7.

    (a) In Fig. 3.7a, pollutant is fully mixed in the whole room (this is the common case for the buffer room).

    (b) In Fig. 3.7b, pollutant is distributed in part of the room, but it does not reach the exit of Room 2 (For Room 3 without buffer room belongs to this case).

    (c) In Fig. 3.7c, pollutant is also distributed in part of the room, but it also reaches the exit of Room 2 (When the room is spacious with shallow depth, this situation may appear).

How is the pollutant distributed, i.e., how is the performance of mixture? It can be expressed with the mixture coefficient α. For small room such as the buffer room, the performance of mixture is good, which is shown in Fig. 3.7a. In this case, α_2 is set 1. For the case with poor performance of mixture, α_2 may be 0.2. When there is no air supply in this case, the mixture coefficient could chosen with half at the maximum, i.e., 0.5. On average, it could set 0.35.

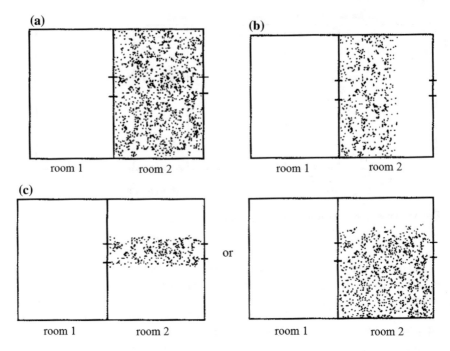

Fig. 3.7 Extent of mixture

Therefore, the resultant concentration can be expressed as follows:

(1) After the door of Room 1 is closed, the initial pollutant concentration in Room 2 is:

$$N_{20} = \frac{N_1 Q_1}{V_2 \alpha_2} \ (\mathrm{pc/m^3})$$

(2) With the self-purification effect of the air by HEPA filter installed in return air or supply air pipeline, when there is no pollutant particle source in the buffer room and in the atmosphere, the self-purification for this increased concentration can be expressed as follows based on the instantaneous concentration equation [10].

$$\frac{N_{2t}}{N_{20}} \approx e^{\frac{-nt}{60}} \tag{3.3}$$

Next the derivation of this equation will be introduced.

For a room with air supply and air return (exhaust) system and HEPA filter installed in the supply air or return (exhaust) air pipelines, the instantaneous concentration N_t inside the room can be expressed as:

$$N_t = \frac{60G \times 10^{-3} + M_n(1-S)(1-\eta_n)}{n[1-S(1-\eta_r)]}$$
$$\times \left\{ 1 - \left[1 - \frac{N_0 n[1-S(1-\eta_r)]}{60G \times 10^{-3} + M_n(1-S)(1-\eta_n)} \right] e^{\frac{-nt[1-S(1-\eta_r)]}{60}} \right\} \quad (3.4)$$

where G is the particle generation rate per volume from occupants and surfaces indoors, pc/(m$^3 \cdot$ min); M_n is the atmospheric particle concentration, pc/L; S is the ratio of the return air volume to the supply air volume; η_n is the total efficiency of filters installed on fresh air pipeline; η_r is the total efficiency of filters installed on return air pipeline; N_0 is the initial concentration indoors.

Let $\frac{60G \times 10^{-3} + M_n(1-S)(1-\eta_n)}{n[1-S(1-\eta_r)]} = A$, Eq. (3.4) can be re-written as:

$$N_t = A \times \left\{ 1 - \left[1 - \frac{N_0}{A} \right] e^{\frac{-nt[1-S(1-\eta_r)]}{60}} \right\} = A - Ae^{\frac{-nt[1-S(1-\eta_r)]}{60}} + N_0 e^{\frac{-nt[1-S(1-\eta_r)]}{60}}$$

For the special case when pathogenic bacteria is released from patients in the isolation ward, it will enter into the buffer room because of the door opening, then it will enter into other rooms. Since there is no bacteria release source in the buffer room and the next following rooms, $G = 0$. Since there is no such bacteria in the atmosphere, $M_n = 0$. So $A = 0$. Therefore, we obtained

$$N_t = 0 - 0 + N_0 e^{\frac{-nt[1-S(1-\eta_r)]}{60}} = N_0 e^{\frac{-nt[1-S(1-\eta_r)]}{60}}$$

Because η_r is the total efficiency of filters installed on return (exhaust) air pipeline, according to Chinese standard, HEPA filters with Type B or higher requirement will be used for this application. The filtration efficiency for bacteria reached more than 99.9999% (refer to Sect. 3.4). So $1 - \eta_r \approx 0$. Therefore, we obtain:

$$N_t = N_0 e^{-nt/60}$$
$$\frac{N_t}{N_0} = e^{(-nt/60)} \quad (3.5)$$

Neither the common cleanroom where there is bacteria generation inside nor the common environment where particles or bacteria exist in the atmosphere is suitable to adopt this equation.

As for Eq. (3.5), when $t \rightarrow \infty$, we obtain

$$\frac{N_t}{N_0} = 0$$

This means $N_t = 0$.

In physics, when a certain amount of bacteria enter into the buffer room during the door opening process, it will be self-purified continuously in the buffer room or diluted by the incoming flow and then exhausted. When $t \rightarrow \infty$, all the bacteria will be captured by HEPA filter or exhausted outdoors, so that air cleanliness level will be recovered in the buffer room.

When $t = \infty$, we obtain

$$\frac{N_t}{N_0} = 1$$

This means $N_t = N_0$. At the moment when bacteria enter into the buffer room, the instantaneous concentration is N_1, which is the initial concentration of the bacteria in the buffer room.

From the above analysis, before the pollutant enters into Room 3, the pollutant concentration in Room 2 (refer to Fig. 3.8) will be

$$N_{2t} = \frac{N_1 Q_1}{V_2 \alpha_2} e^{(-nt/60)}$$

where t is the self-purification time, min. It starts when the door in Room 1 is closed. It equals to the period for the walking from the door in Room 1 to the door in Room 2 before which is open (the interlock time of the door is included). It is usually between 5 s (~ 0.1 min) and 30 s (~ 0.5 min). n is the air change rate, h^{-1}.

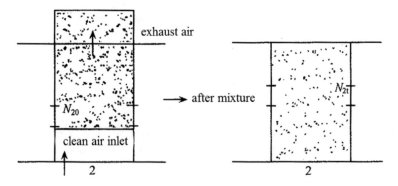

Fig. 3.8 Distribution of pollutant after it enters into room 2

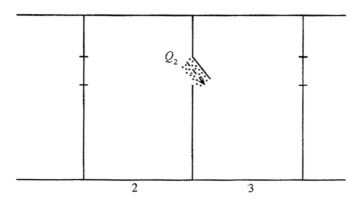

Fig. 3.9 Schematic of pollutant entering into room 3

(3)   When the door of Room 2 is open and occupant moves from Room 2 to Room 3, the pollutant flux at the gate of Room 3 is shown in Fig. 3.9, which is

$$\frac{N_1 Q_1 Q_2}{V_2 \alpha_2} \mathrm{e}^{(-nt/60)} (\mathrm{pc})$$

So when the door of Room 2 is closed, the initial concentration of Room 3 becomes

$$N_{30} = \frac{N_1 Q_1 Q_2}{V_2 V_3 \alpha_2 \alpha_3} \mathrm{e}^{(-nt/60)}$$

After Room 3 is self-purified, the concentration of the airflow entering into Room 4 becomes

$$N_{3t} = \frac{N_1 Q_1 Q_2}{V_2 V_3 \alpha_2 \alpha_3} \left[\mathrm{e}^{(-nt/60)} \right]^2 (\mathrm{pc/m^3})$$

For Room 2, n is big and t is small. While for Room 3, n is small and t is big. For the simplification of assumption in calculation, values of nt are assumed the same, which is shown in Fig. 3.10.

(4)   At the moment when the door of Room 3 is open, the pollutant flux at the gate of Room 4 (buffer room) is shown in Fig. 3.11, which is

$$\frac{N_1 Q_1 Q_2 Q_3}{V_2 V_3 V_4 \alpha_2 \alpha_3 \alpha_4} \left[\mathrm{e}^{(-nt/60)} \right]^2$$

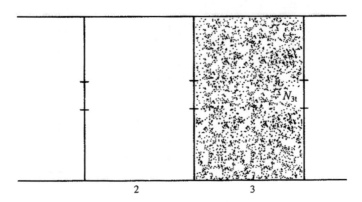

Fig. 3.10 Distribution of pollutant after it enters into room 3

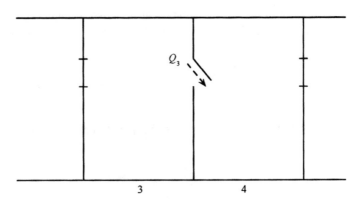

Fig. 3.11 Schematic of pollutant entering into room 4

(5) With the same analysis method for (2), after the pollutant is fully mixed in Room 4, the concentration in Room 4 before the pollutant enters into Room 5, which is shown in Fig. 3.12, becomes

$$N_{4t} = \frac{N_1 Q_1 Q_2 Q_3}{V_2 V_3 V_4 \alpha_2 \alpha_3 \alpha_4} \left[e^{(-nt/60)} \right]^3 (pc/m^3)$$

(6) When the pollutant enters into Room 5 from Room 4, the pollutant flux at the gate of Room 5 becomes

$$\frac{N_1 Q_1 Q_2 Q_3 Q_4}{V_2 V_3 V_4 \alpha_2 \alpha_3 \alpha_4} \left[e^{(-nt/60)} \right]^3$$

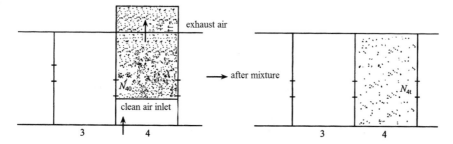

Fig. 3.12 Distribution of pollutant after it enters into room 4

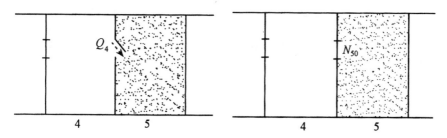

Fig. 3.13 Distribution of pollutant after it enters into room 5

After the door of Room 4 is closed, the initial concentration of Room 5 (refer to Fig. 3.13) becomes

$$N_{50} = \frac{N_1 Q_1 Q_2 Q_3 Q_4}{V_2 V_3 V_4 V_5 \alpha_2 \alpha_3 \alpha_4 \alpha_5} \left[e^{(-nt/60)} \right]^3 (\text{pc}/\text{m}^3)$$

3.2.2   Isolation Coefficient of Buffer Room

1. Isolation coefficient

   Isolation coefficient is the ratio of the initial concentration in the isolation ward to the induced pollutant concentration in the third room by opening of two doors when the buffer room is set. The larger the total isolation coefficient is, the stronger the prevention ability is. It represents the enhancement ratio of the total prevention ability with buffer room to that without buffer room, which is labeled with β.

Suppose $\alpha_2 = \alpha_3 = \alpha_4 = \alpha_5 = \alpha$ for Three-Room-One-Buffer scheme, we obtain

$$\beta_{3\cdot1} = \frac{N_1}{N_{30}} = \frac{V_2 V_3 \alpha^2}{Q_1 Q_2 [e^{(-nt/60)}]} \tag{3.6}$$

When it is the same assumption for Five-Room-Two-Buffer scheme, we obtain

$$\beta_{5\cdot2} = \frac{N_1}{N_{50}} = \frac{V_2 V_3 V_4 V_5 \alpha^4}{Q_1 Q_2 Q_3 Q_4 [e^{(-nt/60)}]^3} \tag{3.7}$$

With the same amount Q (m^3) of incoming flow, suppose the number of the buffer rooms is m and the total number of the rooms is k. Volumes of two isolation wards are usually the same, i.e., $V_3 = V_5 = V$, m^3. Volumes of two buffer rooms are also usually the same, i.e., $V_2 = V_4$, m^3. Suppose the parameter X represents the ratio of the volumes between the isolation ward and the buffer room, we obtain the following general formula.

$$\beta_{k\cdot m} = \frac{V^{(k-1)} \alpha^{(k-1)}}{X^m Q^{(k-1)} [e^{(-nt/60)}]^{(k-2)}} \tag{3.8}$$

2. Example

Calculation of the entrainment airflow Q from one room to the other can be performed as follows. Since the flow rate induced by the pressure difference 5 Pa is less than 0.05 m^3/s, the counteracting effect of the flow rate by the pressure difference less than 5 Pa can be ignored.

From Table 2.11, the flow rate Q of the convection by door opening with $\Delta t = 1$ °C within 2 s is 0.44 m^3.

From Sect. 2.2.5, the maximum flow rate Q of the entrainment flow during the door opening process is 0.9 m^3.

The flow rate of the entrainment flow by occupant movement within 2 s is 0.28 m^3.

Therefore, the total flow rate Q is $\Sigma Q = 1.62$ m^3. (In literatures [7–9], it was 1.52 m^3.)

Suppose $nt = 6$. Since in the buffer room n is very large, we can assume $\alpha_2 = 1$. In Room 3 the performance of the air change with air cleaning technique is good, but the value of n is smaller than that in the buffer room. We can assume $\alpha_3 = 0.8$. On average $\alpha = 0.9$. Suppose the volume of the isolation ward is 25 m^3 and $X = 5$, we obtain

$$\beta_{3\cdot1} = \frac{25^2 \times 0.9^2}{5 \times 1.62^2 \times 0.9} = 42.9$$

$$\beta_{5\cdot2} = \frac{25^4 \times 0.9^4}{5^2 \times 1.62^4 \times 0.9^3} = 2042$$

When Room 4 is the buffer room with positive pressure, we assume that the relative pressure difference between Room 4 and the exterior room is $\Delta P = +5$ Pa. When the non-air-tight door is used, the pressurized outward flow rate Q_4 is

$$Q_4 = 1.62 + 0.08 = 1.7\,\mathrm{m}^3$$

$$\beta_{5\cdot2} = \frac{25^4 \times 0.9^4}{5^2 \times 1.62^3 \times 1.7 \times 0.9^3} = 1943$$

It is shown that the influence of Room 4, whether it is the negative or the positive pressure buffer room, on the result is not large.

The aforementioned isolation coefficients are related to some prescribed parameters such as the size of the door opening, the period of entrainment by occupant, and the induced flow rate. Therefore only the relationship of the order of the magnitude is reflected in the result. Please do not focus on the specific value.

3.2.3 Influencing Factors for Performance of Buffer Room

1. Air change rate

The size of the buffer room is very small. Even when the air change rate reaches 60 h^{-1}, the corresponding flow rate is only 300 m^3/h, which is inconsiderable. Therefore air must be supplied into the buffer room.

For the common range of the air change rate in the buffer room, the influence of the air change rate on the isolation performance is not large.

Figure 3.14 shows the influence of the air change rate on the self-purification of the particle concentration (the background concentration is not included) entering into the buffer room, when the air change rate is within the common range [9].

After the door is closed, it usually takes 0.1 min to walk to the door of the other side. In Eq. (3.5), $e^{-nt/60} = N_t/N_0$. It is shown from Fig. 3.14 that the value with 120 h^{-1} is less than that with 60 h^{-1} by about 10%, and the influence on $\beta_{3\cdot1}$ is slightly more than 10%.

Table 3.3 illustrates the influence of the air change rate on the isolation coefficient. It is shown that when the air change rate increases by one time, the isolation performance only increases by more than 10%.

Of course, when n reaches 1200 h^{-1}, the performance will be improved obviously. However, this is impractical and unrealistic.

2. Volume

The larger the volume of the buffer room is, the better the isolation performance is.

From the general formula mentioned above, with the given air change rate, when the volume of the buffer room is larger and when x is smaller, the isolation

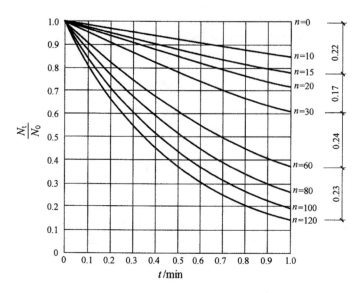

Fig. 3.14 Influence of the air change rate on the self-purification of the particle concentration

Table 3.3 Influence of the air change rate on the isolation coefficient

Air change rate, h^{-1}	t, min	$\beta_{3.1}$
60	0.1	42.9
80	0.1	43.9
100	0.1	45.4
120	0.1	47.7

performance will be better. But when the volume is large, the mixture performance within a certain period will be poor, which means α will be smaller. When the reduction rate of α is not proportional to the increase rate of x, the total isolation performance will improve more. Considering the feasibility for the layout, the volume of the buffer room should not be smaller than 2–3 m^2.

3. Self-purification time

The more the time it takes for occupant to walk through the buffer room, or the more the time it is set for self-lock, the better the isolation performance is.

It is much economic and simpler to increase the self-purification time than to increase the air change rate. When t increases from 6 to 12 s, it is relative easy. But it is equivalent to the increase of the air change rate by one time. Of course, the time for self-lock should not be too long, which is usually within 30 s. Therefore, when serious condition of pollution occurs, the time to open the other door should be delayed, or the time of self-lock should be prolonged to 30 s. Compare with the condition of 6 s, for the air change rate 60 h^{-1} as shown in Fig. 3.28, the value of $e^{-nt/60}$ decreases from 0.9 for the self-lock time 6 s to 0.6 for the self-lock time 30 s.

The value of $\beta_{3.1}$ will increase by 1.5 times, and the value of $\beta_{5.2}$ will increase by $(1.5)^3$ times. The total isolation coefficient will be increased as follows:

$$\beta_{3.1} \text{ will increase from } 42.9 \text{ to } 64.4$$
$$\beta_{5.2} \text{ will increase from } 2042 \text{ to } 6892$$

4. Conclusion

From the above-mentioned points 2 and 3, we could not obtain the conclusion that the smaller the buffer room is, the better the isolation performance will be. With the given air change rate, the volume of the buffer room exerts no influence on the air change rate. But when the buffer room becomes small, the value of x in Eq. (3.7) will become large, and the isolation performance will be poor. When the air supply rate is fixed, with the decrease of the volume of the buffer room, the air change rate will become large, and the isolation performance depends on the relative influences on x and n. Even when the isolation performance increases, in essence it is caused by the increase of the air change rate. It should be noted that if the buffer room is as small as half step between two doors, the value of t may be smaller than 6 s by 50%. In this case, the loss outweighs the gain for improving the isolation performance.

3.2.4 Experimental Validation

1. Experimental scheme

(1) Microbial experiment

In the project entitle with "*Research on isolation performance of the isolation ward*", studies on performance of the buffer room was carried out in Institute of Building Environment and Energy, China Academy of Building Research, Beijing, China. Microbial experiment was performed in a simulated isolation ward, where the aforementioned experiment on the pressure difference was performed. Figure 3.15 shows the appearance of the isolation ward. Figure 3.16 shows the preparatory work performed in the ward. Figure 3.17 illustrates the layout of the ward. Parameters in the experiment are shown in Table 3.4.

Figure 3.18 shows the bacteria solution atomization system. Figure 3.19 shows the simulated bacteria generation condition at the mouth.

In order to check the function of the buffer room, occupant was required to move outwardly through the door by opening and closing the door for one time (about 2 s).

The type of the bacteria generated is the Bacillus atrophaeus (formerly *Bacillus subtilis* var. niger) with the strain number ATCC: 15442; 1.3343. It was provided by Biological Resource Center, Institute of Microbiology Chinese Academy of Sciences (IMCAS-BRC).

Fig. 3.15 Appearance of the isolation ward

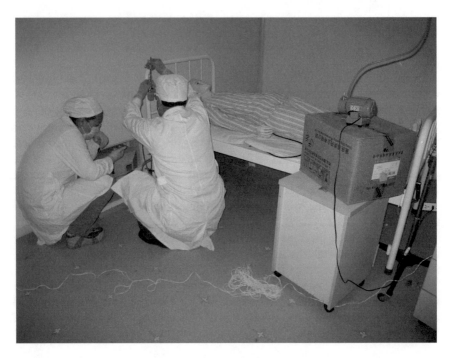

Fig. 3.16 The preparatory work performed in the ward

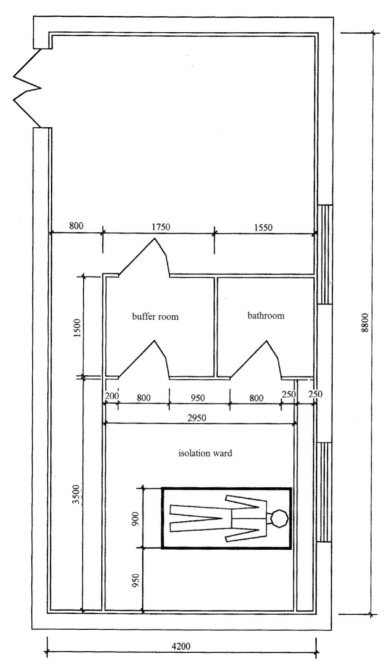

Fig. 3.17 Layout of the simulated isolation ward (in the figure, the bathroom did not work. There is no supply air and exhaust air. If the bathroom were real, the door of the bathroom should be open towards the ward inside)

Table 3.4 Related parameters for determining the isolation coefficient in the buffer room with microbial experiment

Volume		Condition of air exchange	Scheme 1		Scheme 2	
			Temperature (°C)	Pressure difference (Pa)	Temperature (°C)	Pressure difference (Pa)
Isolation ward a	27.6 m³	All fresh air with air change rate 12 h⁻¹	20.1	$\Delta P_{a-b} = -5$	20.2	$\Delta P_{a-b} = 0$
Buffer room b	6.25 m³	No air exhaust. The air supply rate is very small, which is equivalent to the natural ventilation	18	$\Delta P_{b-c} = -5$	18	$\Delta P_{b-c} = 0$
Exterior room c	About 27.6 m³	No air supply and exhaust	Normal temperature	Normal pressure	Normal temperature	Normal pressure

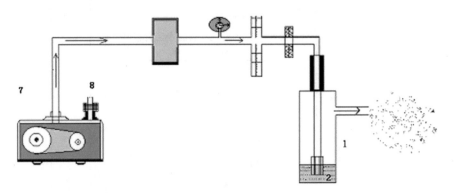

Fig. 3.18 Schematic of the bacteria solution atomization system

There are different opinions in literatures about the naked size of the bacteria, which includes 0.5, 0.8, 1 and 1.5 μm. According to the data in literature [11] published in 2003, the linear length and the width of this kind of *Bacillus subtilis* are 1 and 0.5 μm, respectively. But according to the SEM figure of this paper (which are shown in Figs. 3.20 and 3.21), the linear length for most of the *Bacillus subtilis* is about 1.2 μm, and a few are smaller than 0.5 μm. As for whether the size of the spores generated in our experiment was the same as that in the published literatures, validation was not performed.

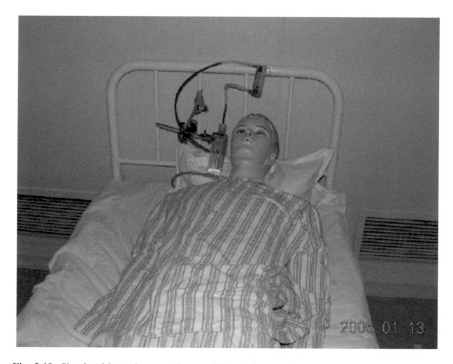

Fig. 3.19 Simulated bacteria generation condition at the mouth

Fig. 3.20 SEM figure of the *Bacillus atrophaeus* (amplification ratio 1700)

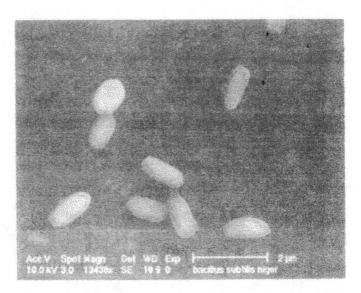

Fig. 3.21 Enlarged SEM figure of the *Bacillus atrophaeus* (amplification ratio 13500)

After being cultured to be colony forming unit, the color of this kind of bacteria becomes light yellow to red, which is rare for the hybrid strain (including *Bacillus subtilis*) in atmosphere. So we could consider the background of this kind of bacteria was zero. Error could be avoided. They are easily discovered. The concentration of the bacteria solution, the quantity of the liquid atomized, and the period used for atomization of the liquid should be controlled, so that the bacteria concentration was not too high to be counted in the isolation ward. At the same time, those bacteria passing through the buffer room can be sampled in exterior room, so that the isolation coefficient could be calculated.

Therefore, according to the trial experiment, the bacteria solution concentration was set 10^{10}–10^{11} pc/mL. In practice the bacteria solution concentration reached 8×10^{10} pc/mL. The bacteria solution concentration was determined by the gradual dilution method with the dilution ratio 10.

According to the microbiological experiment method specified in foreign standard, the quantity of the solution atomized should be between 5 and 10 mL. The quantity of air flow rate for generation of bacteria should be 17 L/min. The period for generation of bacteria should be 30 min.

Based on experience, the generated liquid droplet by the atomizer was between 1 and 5 μm. The pressure for atomization was 1 kg/cm^2.

Bacteria were sampled with the sedimentation method.

The petri dishes should be placed on the floor and near the gate. The measurement points are shown in Fig. 3.22.

(2) Experiment with atmospheric dust

Table 3.5 shows the experiment parameters.

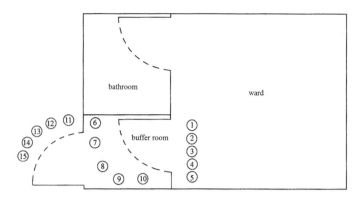

Fig. 3.22 Layout of five sampling points on the floor (the direction of the door is illustrated in Fig. 7.15)

Table 3.5 Related parameters for experiment on isolation coefficient of buffer room with atmospheric dust

Volume		Condition of air exchange	One people open and then close the doors to walk from the ward to the buffer room	
			Temperature, °C	Relative humidity, %
Isolation ward	27.6 m³	All fresh air	20.0	55
Buffer room	6.25 m³	Air supply and exhaust with air change rate 112 h⁻¹	20.6	54
Exterior room	About 27.6 m³	No ventilation	–	–

All fresh air was supplied into the isolation ward when the fresh air did not pass through HEPA filter. When the indoor concentration was close to that of the atmospheric dust outdoors, the indoor concentration was measured. The self-purification process was turned on in the buffer room. Measurement was performed in the buffer room to determine if the air cleanliness level ISO 6 has been arrived.

Then the opening and closing of doors were completed. The concentration in the buffer room was measured immediately. The particles entering into the buffer room were considered as microbes. Measurement was performed every 1 min. Three to four readings were recorded. The concentration in the isolation ward was monitored until it reached stable.

(3)   Experiment with temperature difference

Table 3.6 shows the experiment with the temperature difference. For detailed information about the method, please refer to literature [12].

Table 3.6 Conditions in the experiment with the temperature difference

Condition	Relative pressure difference between the isolation ward and the buffer room, Pa		Relative pressure difference between the buffer room and the exterior room, Pa		Temperature in the ward, °C	Temperature in the buffer room, °C	Relative temperature difference between the ward and the buffer room, °C
	Before opening the door of the ward	During opening the door of the ward	Before opening the door of the ward	During opening the door of the ward			
1	−10	0	+1–2	0	17.0	16.8	+0.2
2	−10	0	+1–2	0	20.8	17.8	+3
3	−10	0	+1–2	0	22.7	17.7	+5

2. Experimental results

(1) Microbiology method [11, 13]

Figure 3.23 shows the results on No. 5 microbial sampling point in the isolation ward.

Figure 3.24 shows the results on No. 5 microbial sampling point in the buffer room.

Fig. 3.23 Figure of the sedimentation bacteria in the ward

Fig. 3.24 Figure of the sedimentation bacteria in the buffer room

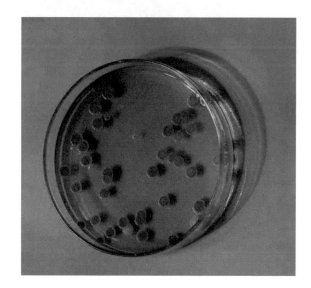

Figure 3.25 shows the results on No. 5 microbial sampling point in the exterior room.

In these figures, the white colony was cultured with the hybrid strain in atmosphere.

It should be noted that only clear figures are presented. Results on No. 5 microbial sampling point in various rooms are relative clear. They cannot be used to obtain the isolation performance exactly, but they can be applied to show the trend of the isolation performance.

Tables 3.7 and 3.8 show the measured data with Scheme 1 and Scheme 2, respectively. In Scheme 1, there is pressure difference and small temperature difference. In Scheme 2, there is no pressure difference and small temperature difference.

Fig. 3.25 Figure of the sedimentation bacteria in the exterior room

Table 3.7 Experimental result on CFU of isolation performance for three-room-one-buffer with pressure difference (−5 Pa) and small temperature difference

Bacteria solution concentration, 8×10^{10} pc/mL
Quantity of the solution atomized, 6.12 mL
Period for liquid spray, 30 min

Isolation ward	Measurement point	1	2	3	4	5	Average
	Data	986	872	823	688	720	817.8
Exterior room	Measurement point	6	7	8	9	10	Average
	Data	–	186	206	160	164	179
Buffer room	Measurement point	11	12	13	14	15	Average
	Data	43	30	40	46	45	40.8

Table 3.8 Experimental result on CFU of isolation performance for three-room-one-buffer with no pressure difference (0 Pa) and small temperature difference

Bacteria solution concentration, 8×10^{10} pc/mL
Quantity of the solution atomized, 6.12 mL
Period for liquid spray, 30 min

Isolation ward	Measurement point	1	2	3	4	5	Average
	Data	800	752	672	784	–	752
Exterior room	Measurement point	6	7	8	9	10	Average
	Data	190	172	194	184	–	185
Buffer room	Measurement point	11	12	13	14	15	Average
	Data	42	40	–	44	–	42

According to the above measured data, the isolation coefficients can be obtained as follows.

For Scheme 1 with pressure difference and small temperature difference,

$$\beta_{3\cdot1} = \frac{817.8}{40.8} = 20.04$$

For Scheme 2 with no pressure difference and small temperature difference,

$$\beta_{3\cdot1} = \frac{752}{42} = 17.9$$
$$\beta_{3\cdot1} = 19$$

(2) Experimental method with atmospheric dust [14]

The isolation performance with atmospheric dust is obtained through proceeding of data in Table 4.4, which is shown in Table 3.9.

(3) Experimental method with temperature difference

It has been shown in Chapter 2.

(4) Experimental method with artificial dust

From the data provided in literature [15], the isolation performance with artificial dust for the scheme with one operating room and the other buffer corridor is obtained, which is shown in Table 3.10.

3. Analysis

(1) Comparison between theoretical analysis and experimental result

Now the isolation performance with the microbiology method will be analyzed. In the buffer room, there is no exhaust air. The air supply volume can only provide

Table 3.9 Experimental result on isolation coefficient for two-room-one-buffer (one isolation ward and one buffer room) with atmospheric dust

Relative pressure difference between the ward and the buffer room, Pa		Isolation coefficient	Note
One people walks out for 2 s	−31	37.0	Maximum concentration appears at the second minute
	−30	40.0	Maximum concentration appears at the first minute
	−6	38.5	Maximum concentration appears at the second minute
	0	24.4	Maximum concentration appears at the first minute
	0	18.9	Maximum concentration appears at the first minute
Average		31.8	

Table 3.10 Experimental result on isolation coefficient for two-room-one-buffer (one operating room and one buffer corridor) with atmospheric dust

Type of door	Relative pressure difference between the operating room and the buffer corridor, Pa	Average isolation coefficient
Outwardly opening door	0	$\frac{2.6\times10^8}{32.4} \div \frac{4.2\times10^6}{13} = 24.2$
	−30	$\frac{2.6\times10^8}{32.4} \div \frac{1.7\times10^6}{13} = 59.8$

Note In the above equations, the values 32.4 and 13 represent the volumes of the operating room and the buffer corridor, respectively

the extremely small amount to maintain the pressure drop across the gap. Therefore, the mixing performance in the buffer room is relatively poor. α_2 should be smaller than 1, which could be set 0.84. There is no ventilation in the exterior room. According to the analysis in previous section, α_3 could be set 0.35.

For the scheme with no pressure difference and small temperature difference, the flow rate of air passing through one room to the other room should be as follows. The temperature difference between the buffer room and the exterior room was $\Delta t = 1\ °C$, so the total air exchange rate was $Q = 1.62\ m^3$. The temperature difference between the isolation ward and the buffer room was $\Delta t = 2\ °C$, so the total air exchange rate was $Q = 1.8\ m^3$. According to Table 3.4, we know

$$X = \frac{27.6}{6.25} = 4.42$$

Since in fact there was only air supply with $n < 60$ and the sedimentation samplings were performed simultaneously in the isolation ward and the buffer room, the influence of the time could be ignored. When $t = 2$ s, $nt < 6$ and $e^{(-nt/60)} \approx 1$. Therefore, in theory we obtained

$$\beta_{3\cdot1} = \frac{27.6^2 \times 0.85 \times 0.35}{4.42 \times 1.62 \times 1.8 \times 1} = 17.6$$

The theoretical value is quite close to the measured data which is $\beta_{3\cdot1} = 17.9$.

For the scheme with pressure difference and small temperature difference, when the flow rate (in 2 s) 0.082 m^3 by negative pressure is counteracted, in theory we obtain $\beta_{3\cdot1} = 18.5$. It is also close to the measured value 20.04.

(2) By experiments on isolation coefficient with the above-mentioned microbiology method and the experimental method with temperature difference, the isolation coefficients with the microbiology method and the temperature difference method are much smaller than that with the atmospheric dust in the scheme with one isolation ward and one buffer room. The experimental value of the isolation coefficient with atmospheric dust is much larger than the theoretical value. The reason is the self-purification time, which will be analyzed as follows.

The theoretical isolation coefficient is defined by Eq. (3.9), which only considers the entrance of "exterior particles" from the isolation ward to the buffer room.

$$\beta_{2\cdot1} = \frac{N_1}{N_{2t}} = \frac{N_1}{\frac{N_1 Q_1}{V_2 \alpha_2} e^{-nt/60}} = \frac{V_2 \alpha_2}{Q_1 e^{-nt/60}} \tag{3.9}$$

For the data in Table 3.4, the related parameters are as follows.
$V_2 = 6.25$ m^3
$\alpha_2 = 1$ (with exhaust air)
$Q_2 = 1.62$ m^3/s
$n = 112.5$ h^{-1}

The maximum value for the pressure difference 0 Pa appears at 1 min (which is shown in Table 3.7), so $t = 1$ and we obtain
$e^{-nt/60} = 0.153$

Therefore, for the pressure difference 0 Pa, the isolation coefficient is

$$\beta_{2\cdot1} = \frac{6.25 \times 1}{1.62 \times 0.153} = 25.2$$

The theoretical value 25.2 under the real situation is very close to the average measured data 21.7 shown in Table 3.7.

The maximum value for the pressure difference -30 Pa also appears at 1 min, so $t = 1$. But for the case with large pressure difference, the flow rate by the negative pressure should be counteracted. From Table 2.1, $Q \neq 1.62$. Since the air velocity through the door gate is 0.112 m/s, with the area of the door we obtain the flow rate

0.22 m³/s. So $Q = 1.4$. Therefore, for the pressure difference -30 Pa, we obtain the isolation coefficient

$$\beta_{2\cdot1} = \frac{6.25 \times 1}{1.4 \times 0.153} = 29.2$$

It is very close to the experimental data 40.

The maximum value for the pressure difference -31 Pa also appears at 2 min, so $t = 2$ and we obtaine$^{-nt/60} = 0.025$

Therefore, for the pressure difference -31 Pa, the isolation coefficient is

$$\beta_{2\cdot1} = \frac{6.25 \times 1}{1.4 \times 0.025} = 178.6$$

The theoretical value 178.6 under the real situation is far from the average measured data 37. The reason may be that for the timing of the optical particle counter, when the end of the reading was 2 min, which may be in fact just passing through 1 min. But it was recorded in the region of 2 min. When 1 min was used for calculation, the theoretical value $\beta_{2\cdot1}$ became 29.2, which is close to the measured data 37.

In literature [16], the experimental data was also calculated with Eq. (3.9) by the Japanese scholar, i.e.,

$$\beta_{2\cdot1} = \frac{V_2 \alpha_2}{Q_1 e^{-nt/60}}$$

where V_2 is the volume of the isolation corridor, which could set 13 m³.

α_2 is the coefficient, which could set 1 for the buffer room where air is supplied and exhausted.

Q_1 is the flow rate. The size of the door is the same as that of the previous case. The time for the previous case is 2 s, but in this case it is 1.6 s. When people arrive at the door, they began to open and close the door. The average time of this process for ten people was obtained. The average flow rate obtained was 1.3 m³ (not 1.62 m³).

n is the air change rate, which is $260/13 = 20$ h^{-1}.

t is the first minute when the maximum concentration on the reading of the optical particle counter.

Therefore, for the pressure difference 0 Pa, the isolation coefficient is

$$\beta_{2\cdot1} = \frac{13 \times 1}{1.3 \times 0.72} = 14$$

For the case with the pressure difference -30 Pa, the flow rate by the negative pressure should be counteracted. $Q = 1.1$ m³. The theoretical value $\beta_{2\cdot1}$ became 16.4.

(3) During the experimental method with temperature difference, the air supply rate in the buffer room is 379 m³/h and $n = 60.64$ h^{-1}. Since there is air supply and air exhaust with large air change rate, $\alpha = 1$.

Based on past calculation data, the flow rates Q under different temperature difference conditions is: 1.38 m³ for 0.2 °C, 1.92 m³ for 3 °C, and 2.14 m³ for 5 °C.

In the previous 2 min, $e^{-nt/60} = 0.134$. Theoretical and experimental data for the isolation coefficient are shown in Table 3.11.

It is shown that the calculation result in theory could be used to estimate the actual isolation performance on average.

Both the theoretical and measured data are summarized in Table 3.12. The difference between the theoretical and measured data is large only for the method with atmospheric dust. This is related to the explanation of the original data. But for other methods, the difference is very small. When some parameter is not accurate enough, the influence on the result will be very large. For example, when the error for determining the value of t is several seconds, this influence will appear. Therefore, there is still relative reference value in the theoretical equation.

Table 3.11 Theoretical and experimental data for the isolation coefficient of the outwardly opening door during the experiment with the temperature difference

Δt, °C		0.2	3	5
$\beta_{2\cdot1}$	Calculation	33.9	24.3	21.9
	Experiment	28.6	23.2	15.2

Table 3.12 Comparison of the theoretical and measured isolation coefficients

Type of experiment	Country	Constitution of rooms	Theoretical value	Measured data
Microbiology method	China	Three-room-one-buffer	17.6 for 0 Pa 18.5 for 5 Pa	17.9 20.04
Method with atmospheric dust	China	Two-room-one-buffer	25.2 for 0 Pa 29.2 for 30 Pa 178.6 or 29.2 for 31 Pa	21.7 40.0 37.0
Method with artificial dust	Japan	Two-room-one-buffer	14 for 0 Pa 16.4 for 30 Pa	24.2 59.8
Method with temperature difference	China	Two-room-one-buffer	16.9 (average value with three data under different temperature difference values)	22.3 (average value with three data under different temperature difference values)

Table 3.13 Influence of the pressure difference on the isolation performance

Type of experiment	Amplification ratio of the isolation performance in experiment	
	Ratio for −5 (or −6) Pa to 0 Pa	Ratio for −30 to 0 Pa
Microbiology experiment	0.12	−
Method with atmospheric dust	0.67	0.67
Method with artificial dust	−0.7	1.47

(4) Influence of the pressure difference on the isolation performance

Based on the aforementioned experimental data, it is shown that the influence of the pressure difference on the isolation performance is not large. This is because the counteracting flow rate by the negative pressure difference on the outward leakage flow rate is very small. Results are shown in Table 3.13.

It takes too much effort to create the pressure difference −30 Pa, compared with the situation with the pressure difference 0 Pa. However, the amplification magnitude for the increase of the isolation coefficient is not too much.

Besides under the condition with the pressure difference −30 Pa, when there is temperature difference, the pollutant cannot be prevented from leakage outwardly during the opening process of the door. From Table 2.3, the quantity of the leakage flow rate still reached more than one fifth of the original concentration. Therefore, the cost-effective performance to adopt the pressure difference −30 Pa is almost the same as that of the situation with −5 Pa.

In short, it is not reasonable and safe to pursue the isolation performance by increasing the pressure difference. When the pressure difference increases from 0 to −30 Pa, the isolation performance is increased by more than one time. However, when one buffer room is added, the isolation coefficient $\beta_{3.1}$ will be increased by a dozen times.

3.2.5   Door of Buffer Room

There are two doors for the buffer room. One is the door adjacent to the isolation ward, which can also be called the door of the ward. The other is the door for entrance into the interior corridor. For the purpose of convenience, the former door is called the inner door of the buffer room, and the latter is called the outer door of the buffer room.

As mentioned before, the extent of the air-proof for the inner door has no effect on the outward leakage of air flow. However, there are still differences for different kinds of the doors. From Fig. 2.6 and Table 2.8, the performance of the outwardly opening door is poorer than that of the inwardly opening door which has worse performance than the sliding door.

The air velocity of the counter current has been introduced before. Figure 3.26 illustrates the air velocities of the counter current during the opening and closing of three different door under different pressure difference conditions. In the figure, the component of $-x$ means the direction from indoors towards outdoors, and the component of $+y$ means the direction from the heel post of the door towards the door handle, which are shown in Fig. 3.27. Since the air velocity of the y component during the closing of the door is larger than that during the opening of the door, the air velocity of the $+y$ component during the closing of the door is presented in the figure [15].

In this experiment, the air velocity of the counter current reached more than 1 m/s, which is obviously larger than the measured values by American scholar and our study. In this experiment, the air velocity is the one during the pushing process

Fig. 3.26 Relationship between the air velocities of the counter current during the opening and closing of the door and the pressure difference

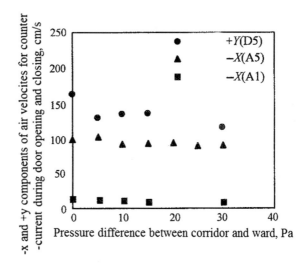

Fig. 3.27 Schematic diagram of the air velocity components for the counter current

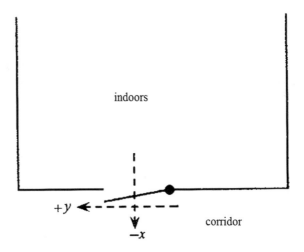

of the door, while in our study the air velocity was measured after the door was open. This may be the reason for the difference.

From the figure, it is shown that:

(1) The magnitude of the air velocity for the counter current is: outwardly opening door > inwardly opening door > sliding door. This can be used to explain the sequence for the relationship between the particle number transmitted during the opening of the door and the type of the door shown in Fig. 2.6.

(2) The air velocity of the counter current is not related to the pressure difference. This means the air velocity of the counter current for $\Delta P = 0$ Pa is close to that for $\Delta P = 30$ Pa. It even beyonds the imagination that for the inwardly opening door, the air velocity of $\Delta P = 0$ Pa is the maximum while that for $\Delta P = 30$ Pa is the minimum.

From the aforementioned analysis, with the suitable size of the buffer room, the sliding door could be adopted as the inner door, and the outwardly opening door could be utilized as the outer door. For the ward with negative pressure, the outwardly opening door can be used only. The problem of increasing the air-proof performance cannot be considered. The door with common air-proof performance is fine, except that the wooden material is not used.

3.3 Airflow Isolation in Mainstream Area

3.3.1 Concept of Mainstream Area

Medical personnel for ward round or operation near the sickbed will face the pollutant source directly. During the talking, coughing and sneezing processes of patients, it will pose a threat to the medical personnel. The range hood introduced in Chapter One has been denied by practice. There is a certain effect of prevention for wearing common masks. As for the designer of the isolation ward, how can we reduce the infectious risk of the medical personnel under the dynamic condition of the operation with the current air cleaning system?

For protecting the medical personnel, CDC manual in U.S.A. proposed that clean air should be passing through the working area of the medical personnel. No only CDC emphasized this point, but also other related literatures has mentioned it. But one fact has been ignored, which is shown in Fig. 3.28. When clean air is supplied from the back of the people, negative pressure area will be formed in the front breathing zone of the people, which has no protective effect and is harmful.

Therefore, in the concept of dynamic isolation theory, the measures to utilize the mainstream area are proposed, which has been validated effectively.

In 1979, the concept of the mainstream area was proposed by author [3], which is shown in Fig. 3.29. The region of the downward supply air below the air supply outlet can be termed as the mainstream area. In this area, the air cleanliness level is

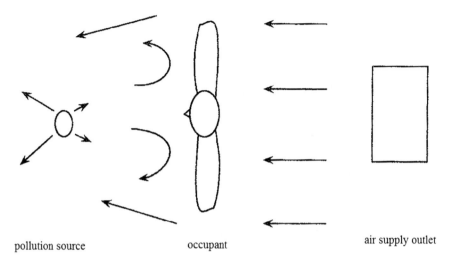

pollution source   occupant   air supply outlet

Fig. 3.28 Negative pressure area formed in the front zone of the people

the best and the ability to exhaust pollution is the best. At the upper region of the
mainstream area, some of the surrounding air will be sucked in, which will be
diluted and then exhausted together from the lower part of the room.

When air is supplied and then exhausted as shown in Fig. 3.29, the average
indoor concentration is:

$$N_v = N_s + \psi \frac{60G \times 10^{-3}}{n}$$

where N_v is the average indoor concentration; N_s is the concentration of the supplied air; G is the bacteria generation rate per unit volume; n is the air change rate;
ψ is the non-uniform distribution coefficient (please refer to Table 3.14).

The parameter ψ can be calculated as follows.

$$\psi = \left(\frac{1}{\varphi} - \frac{\beta}{\varphi} + \frac{\beta}{1+\varphi} \right) \times \left(\varphi + \frac{V_b}{V} \right) \tag{3.10}$$

where φ is the carrier ratio of the airflow. The value φ becomes 1.5, 1.4, 1.3, 0.65,
0.3 and 0.2 when the area corresponding to each air supply unit with air filter inside
on the ceiling is ≥ 7, ≥ 5, ≥ 3, ≥ 2.5, ≥ 2 and ≥ 1 m^2, respectively. β is the
ratio of the particle generation rate in the mainstream area to that in the whole room;
V is the room volume; V_b is the volume of the vortex area.

For the isolation ward with single patient and with the air change rate $n \approx 10^{-1}$,
the value of ψ is 1.5 based on Table 3.14. In this case, the average indoor concentration is:

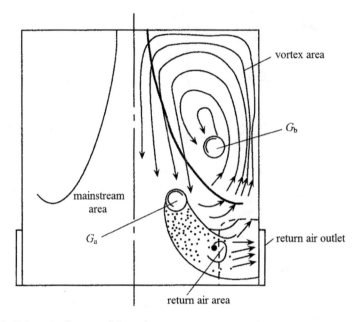

Fig. 3.29 Schematic diagram of the mainstream area

Table 3.14 Non-uniform distribution coefficient ψ	Air change rate, h^{-1}	10	20	40	60	80
	ψ		1.5	1.22	1.16	1.06	0.99

$$N_v = 1.5 \times \left(N_s + \frac{60G \times 10^{-3}}{n} \right)$$

3.3.2 Function of Mainstream Area

The mainstream area was proposed to protect the medical personnel. Air supply outlet is placed on top of the positions where medical personnel performed the ward round and operation near the sickbed of the patient, which is shown in Fig. 3.30.

If there is no particle generation source in the mainstream area as shown in Fig. 3.30, $\beta = 0$. The area of the room is larger than 10 m^2. If there is only one air supply outlet, the parameter is $\varphi = 1.5$.

Therefore, the non-uniform distribution coefficient in the mainstream area becomes

$$\psi = 1 - \frac{\beta}{1 + \varphi} = 1 - \frac{0}{1 + 1.5} = 1$$

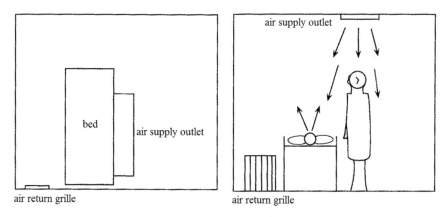

Fig. 3.30 Schematic of the medical personnel near the sickbed under the protection of the mainstream area

The average concentration in the mainstream area is

$$N_a = N_s + \frac{60G \times 10^{-3}}{n}$$

Therefore, based on the expressions of the concentration in the mainstream area and the average indoor concentration N_v, we obtain

$$N_a = 0.67 N_v$$

This means that the pollutant concentration in the mainstream area where the medical personnel stay and face as the breathing zone is about 2/3 of the average indoor concentration. This is much less than that in the vortex area.

In other words, when the medical personnel stays in any place indoors (except the breathing area in front of the patient), the pollutant concentration could be considered as the average indoor concentration N_v. But when the medical personnel stay inside the mainstream area, the pollutant concentration can be reduced to 2/3 of N_v.

Compared with the concentration in the breathing zone of the patient, the concentration in the mainstream area is much smaller. This is because the pollutant entering into the mainstream area only accounts for a small proportion, which is further diluted by the supplied air in the mainstream area. Therefore, the concentration of the mainstream area is much less than 2/3 of the average indoor concentration.

Because in the ward, patient stays inside alone, or stays with other patients who carry the same kind of disease. They are already adapted to this habitat environment. The average indoor concentration is not too important for them. However, for the medical personnel, although they walk around in the ward, they will usually stay for a long period besides the sickbed. Because they are healthy people, they are sensitive to the indoor pollutant. Therefore, the pollutant concentration in this region should be as small as possible.

This is the reason why the air supply outlet should be placed above the side of the sickbed where the medical personnel usually stay. It is aimed to protect the medical personnel through the mainstream area.

Because of such performance of the mainstream area, the mainstream area should be enlarged as much as possible. With the given area of the air filters, the area of the air supply outlet should be increased, as long as the air velocity at the air supply outlet is not small than 0.13 m/s [4]. Otherwise, the performance will be opposite.

3.4 Application of Self-circulation Air Through HEPA Filter

3.4.1 Application Principle of Circulation Air

From Chap. 2, we know that it is a misunderstanding that circulation air is not allowed for usage. But for the application of circulation air in the concept of the dynamic isolation, there are several principles:

(1) The circulation air does not mean the circulation of the return air in the central system for different rooms. It is the self-circulation for the air in each room.
(2) There must be a proportion of circulation air to be exhausted outdoors, so that negative pressure can be kept indoors. On the exhaust air outlet, HEPA filter must be installed.
(3) The exhaust air outlet and the return air outlet can be combined to be one apparatus. HEPA filter could be shared for both the exhaust air and the return air.
(4) The exhaust air outlet and the return air outlet are not required to be free of leakage.

3.4.2 Function of HEPA Filter

1. Common disinfection methods

   (1) Features of common disinfection method
       In order to deal with pollution and infection in hospitals, pipeline cleaning and spraying sanitization were usually adopted in the past. Chemical and physical disinfection methods were utilized indoors. This kind of treatment methods with pollution at first and then control is termed as the passive treatment. It will yield twice the result with half the effort. Besides, it is a method with sequela. It is not a method to control pollution during the

whole process. Instead, it only pays attention to pollution control at the beginning and the end.

In hospitals the objects where disinfection is needed include: occupant, surfaces of object, and air. For occupant, hands need to be disinfected. People needs to take a bath and wear aseptic clothes.

For surfaces of object, it should be wiped clean. The following methods may be used, which include the disinfection with chemical agent (more than one type must be used), the moist heat sterilization, the dry heat sterilization, the radiation sterilization, the gas sterilization, sterilization with air filtration (when sterilization cannot be performed eventually in the container).

(2) Examples of air disinfection method

Table 3.15 illustrates various methods for air disinfection, where the air disinfection efficiencies were shown in the product catalog, literatures, claim from manufacturer or the test report.

Table 3.16 illustrates the measured result on electrostatic sterilization equipment performed by experts from Institute of Microbiology and Epidemiology at the Academy of Military Medical Sciences [17].

Table 3.17 shows the report from Chinese Journal of Nosocomiology [18].

Table 3.18 shows the data from Tianjin University [19] and National Center for Quality Supervision and Test of Building Engineering [20]. Table 3.19 presents the report data by National Center for Quality Supervision and Test of Building Engineering. Table 3.20 provides the data from the thesis at China Academy of Building Research [20]. Table 3.21 shows the data from literature [21]. Table 3.22 presents data from the Academy of Military Medical Sciences [17] and the dissertation at Tongji University [22].

2. Features of disinfection method with HEPA filter

(1) Whole-process control

When air cleaning system with HEPA filter is used during the whole operation process indoors, the indoor air environment can be controlled. This is different from "sterilization at the beginning" or "sterilization at the end".

(2) Both dust and bacteria are removed

By using the principle of physical barrier, air filter can remove dust, as well as microbes (including bacteria and virus). Since microbes are carried by particles, bacteria can be removed during the removing process of dust.

(3) No generation of other gradients

Some sterilization method may generate toxic gases such as nitrogen oxides or ozone. However, air filtration is a pure physical method, which will not generate other gradients.

(4) No side effect of generating toxic material

Some sterilization method may generate radiation, which is harmful to occupant's health. Some generated electric field or magnetic field has influence on instrument and equipment. Some may promote the mutation of

Table 3.15 Various methods for air disinfection

Disinfection method	Principle	Efficiency
Single-stage electrostatic precipitator	High voltage electric field will form corona, and generate free electron and ion. The dust and bacteria will contact them to become charged. Then they will deposit on the dust collecting electrode to be removed. For relative larger particles and fibers, the efficiency is poorer, because discharge will occur. The advantage is that it's able to remove dust and bacteria with small pressure drop. The shortcoming is that it is difficult and time-consuming to wash, and pre-filter must be installed. It may generate ozone and nitrogen oxides, which forms secondary-contamination problem	50% (test on some products show that the efficiency only reaches about 20%)
Plasma	Under the condition of heating or strong electromagnetic field, gas will generate the electron cloud. Active free radical and rays have broad-spectrum germicidal effect. But it cannot remove dust	66.7%
Atractylodes lancea	Traditional Chinese medicine	68.2%
Anion	With the electric field, UV irradiation field, ray and impact of water, air is ionized to generate anion. It can adsorb dust particle, so that dust becomes heavy ion to settle down. The shortcoming is that secondary airborne dust can be formed. It is of little use in HVAC system	73.4%
Nano-photocatalysis	With the irradiation of sunlight and UV, oxidative decomposition of the volatile organic gas or the bacteria occurs on the surface of catalytic active material. They are converted to CO_2 and water. The disinfected air must contact the catalytic material for a certain period. With the dust loading process, the performance becomes poor. So pre-filter must be installed. UV irradiation will generate ozone. In experiment, the efficiency may become negative	75% (test on some products show that the efficiency only reaches only about 30%, and some of the efficiency even became negative)
Formaldehyde fumigation	It is the chemical agent, which has been confirmed carcinogenic.	77.42%

(continued)

Table 3.15 (continued)

Disinfection method	Principle	Efficiency
Ultraviolet irradiation	The air velocity in HVAC system is very large. When it is applied in HVAC system, the irradiation dose on bacteria is very small, so the performance is poor. Only bacteria can be removed, but dust cannot. Ozone will be generated. WHO and GMP from EU claim that this method is not acceptable. It can not be used as the final disinfection method	82.90%
Electron sterilization lamp	Physical method	85%
Double-stage electrostatic precipitator	The ionization electrode is separate from the dust collecting electrode	90% (test on some products show that the efficiency only reaches about 60%)
Ozone	It is a light blue gas, its oxidation performance is strong. The oxygen atom generated during the oxidation process can oxidize and penetrate through the cell wall of the bacteria. It has broad-spectrum effect of germicide, but it cannot remove dust. During usage, occupant should not stay indoors. Various goods may be destroyed. Little effect will be exerted on surface microorganism. It is detrimental to the respiratory tract of people. This method is not suggested for usage	91.82%
High and medium efficiency air filter with ultra-low resistance	It is a method with physical barrier. When it is used for common air supply outlet, the pressure difference is only about 10 Pa, which is one third of the value for coarse filter. But the efficiency is high or medium (for particles with diameter ≥ 0.5 μm, the efficiency reaches more than 70–80%). It is light and easy for installation, and there is no secondary-contamination	92–98%
HEPA filter	It is a method with physical barrier. There is no side effect. It is disposable. In the "*Technical Standard For disinfection*" issued by Ministry of Health of the People's Republic of China, only air filtration is proposed for air disinfection in cleanroom. The resistance is large	99.9999–99.99999% (test with *Bacillus subtilis* in 2004)

Table 3.16 Bacteria removal efficiency by electrostatic sterilization equipment

Sampling time after start-up, h	Sterilization efficiency	
	Operating room	ICU
0.5	61.20	55.20
1.0	50.40	73.20
1.5	57.00	65.10
2.0	50.80	86.50
3.0	78.80	78.90
4.0	66.70	78.60

Note Bacterial concentration before start-up was 504–900 CFU/m^3

Table 3.17 Comparison on sterilization performance of different disinfection methods

Disinfection method	Bacterial concentration before disinfection, CFU/m^3	Bacterial concentration after disinfection, CFU/m^3	Disinfection rate, %	Note
Atractylodes lancea	642	204	68.22	Experiment was performed by Wu Chun-lan, which was published in Chinese Journal of Nosocomiology
Ozone disinfection machine	673	55	91.82	
Ultraviolet ray	601	103	82.86	
Formaldehyde fumigation	598	135	77.42	

Table 3.18 Data from Tianjin University [19]

UV lamp installed in duct	V = 3 m/s	One-pass sterilization efficiency 80%	Experiment was performed by Wang Ying, which was published in Journal of Tianjin University

Table 3.19 Report data by National Center for Quality Supervision and Test of Building Engineering

Sterilization equipment with circulation-air and UV lamp installation in a room with area 11.6 m^2	Coarse, medium and sub-high efficiency air filters installed in the pipeline of the outdoor air	The sterilization rate indoors after start-up reaches 93%	Experiment was performed by National Center for Quality Supervision and Test of Building Engineering

Table 3.20 The data from the thesis at China Academy of Building Research

(1) One-pass sterilization efficiency of high and medium efficiency air filter with ultra-low resistance

At the half-th minute	At the eighth minute	At the thirteenth minute	At the nineteenth and a half minute	At the twenty-first minute	At the twenty-fifth minute	Average	
99.34%	99.34%	99.39%	99.47%	99.26%	99.43%	99.37%	Experiment was performed by National Center for Quality Supervision and Test of Building Engineeringf building engineering. Authors include Pan Hong-Hong, Cao Guo-Qing, et al.

(2) Pressure drop of high and medium efficiency air filter with ultra-low resistance

Air velocity, m/s	0.31	0.41	0.52	0.61	0.71	0.79	0.89	0.99	Experiment was performed by National Center for Quality Supervision and Test of Building Engineering. Authors include Pan Hong-Hong, Cao Guo-Qing, et al.
Pressure drop, Pa	8	10	12	13	15	17	19	21	

(3) Efficiency of HEPA filtered

Type B HEPA filter based on national standard	Type C HEPA filter based on national standard	Research report from China Academy of Building Research. Authors include Zhang Yi-Zhao, Yu Xi-Hua, et al.
One-pass efficiency is 99.99997% for *B. Subtilis* spores	One-pass efficiency is 99.999997% for *B. Subtilis* spores	

Table 3.21 Efficiency of electrostatic air cleaner

Electrostatic air cleaner	The first ionization (U.K.)	One-pass dust removal efficiency is 80%	*<Fundamentals of air cleaning technology>*
Electrostatic air cleaner	The first ionization (Japan)	One-pass dust removal efficiency is 72.8%	
Electrostatic air cleaner	The secondary ionization (China)	One-pass dust removal efficiency is 99.3%	

Table 3.22 Sterilization efficiency of electrostatic air cleaner

Electrostatic air cleaner	4 h after start-up	Sterilization efficiency in operating room is 66.7%	Report by Yang Ming-Hua by Institute of Microbiology and Epidemiology at the Academy of Military Medical Sciences
Electrostatic air cleaner	0.5 h after start-up	Sterilization efficiency in ICU is 61.2%	Thesis by Mao Hua-Xiong from Tong University
Electrostatic air cleaner	1 h after start-up	Sterilization efficiency with atmospheric bacteria is 79.9%	
	2 h after start-up	Sterilization efficiency with atmospheric bacteria is 91.1%	

bacteria. For example, this will occur during the UV irradiation. The drug resistance of the bacteria may become very strong.

(5) High and complete sterilization efficiency

For HEPA filter, the sterilization efficiency could reach more than 99.99999%. The sterilization performance is complete. There are no semi-lethal bacteria left, which may revive afterwards. For bacteria suffering UV irradiation, they may revive under the light if they are not killed by irradiation. Bacteria corpse or secretion will not left if HEPA filter is used.

(6) Broad-spectrum of sterilization efficiency and easy for selection

The range of sterilization efficiency for different products is from 10 to 99.999999%, while that of other sterilization methods is very narrow, which is from 70 to 90%.

(7) It is not selective for dust and bacteria removal

The performance of some sterilization method such as the electrostatic method is influenced by the conductivity property of the dust particles. Some sterilization method is selective for the type of bacteria, which means the sensitivity level is different.

For example, the UV-C ray with wavelength 253.7 nm has different sensitivity levels for various microbes, which is shown in Table 3.23 [23].

Table 3.23 Sensitivity of various typical microbes for UV ray

Sensitivity to UV ray	Group of microbes	Typical microorganism
The most sensible ↓ ↓ ↓ ↓ ↓ ↓	Endophytic bacterium	*Staphylococcus arueus*
		Streptococcus pyogenes
		Escherichia coli
		Pseudomonas aeruginosa
		Serratia marcescens
	Mycobacteria	*Mycobacterium tuberculosis*
		Mycobacterium bovis
		Mycobacterium leprae
	Spore bacteria	*Bacillus anthracis*
		Bacillus cereus
		Bacillus subtilis
	Fungal spore	*Aspergillus versicolor*
		Penicillium chrysogenum
The least sensible		*Stachybotrys chartarum*

(8) It is a method to prevent aerosol and airborne microbes from entering into the air ducts.

It is equivalent to the method of preventing the enemy from invading into our country. It is the highest state of war to subdue the enemies without fight. This method will not cause corpses of the bacteria cover the plain. Therefore, it is an active pollution control method, and not the passive method to perform disinfection after bacteria come indoors. It is a green method to build the green hospitals. No new pollution will be generated.

Of course, the main shortcoming of air filter is its relative large pressure drop. But pre-filter or filter for air supply outlet should also be installed for some methods, including the nano-photocatalysis and UV equipment. Air filters with lower pressure drop should be utilized. Related technique and products have been available.

1. Classification of air filters

(1) Table 3.24 shows the classification of air filters according to national standard GB/T 14295-2008.

(2) Classification for HEPA and ULPA filter
Classifications for HEPA and ULPA filters according to national standard GB/T13554-2008 are shown in Tables 3.25 and 3.26, respectively.

(3) Comparison of classification standards between China and abroad
The test conditions for air filters in standards from China and abroad are quite different, including the particle source and its diameter. Only approximated instead of quantitative comparison can be performed, which is shown in Fig. 3.26.

Table 3.24 Performance classification of air filters

Type	Symbol	Face velocity, m/s	Efficiency E under rated flow rate, %		Initial pressure drop ΔP_i under rated flow rate, Pa	Final pressure drop ΔP_f under rated flow rate, Pa
Sub-high efficiency	YG	1.0	Particle diameter ≥ 0.5 μm	$99.9 > E \geq 95$	≤ 120	240
High and medium efficiency	GZ	1.5		$95 > E \geq 70$	≤ 100	200
Medium efficiency 1	Z1	2.0		$70 > E \geq 60$	≤ 80	160
Medium efficiency 2	Z2			$60 > E \geq 40$		
Medium efficiency 3	Z3			$40 > E \geq 20$		
Coarse efficiency 1	C1	2.5	Particle diameter ≥ 2.0 μm	$E \geq 50$	≤ 50	100
Coarse efficiency 2	C2			$50 > E \geq 20$		
Coarse efficiency 3	C3		Arrestance with standard artificial dust	$E \geq 50$		
Coarse efficiency 4	C4			$50 > E \geq 10$		

Note When measured efficiency value satisfies two types in this table, the higher classification level can be used for assessment

Table 3.25 Classifications for HEPA and ULPA filters

Classification for HEPA filter

Type	Efficiency E with sodium flame method under rated flow rate, %	Efficiency E with sodium flame method under 20% of rated flow rate, %	Initial pressure drop ΔP_i under rated flow rate, Pa
A	$99.99 > E \geq 99.9$	No requirement	≤ 190
B	$99.999 > E \geq 99.99$	99.99	≤ 220
C	$E \geq 99.999$	99.999	≤ 250

Classification for ULPA filter

Type	Particle counting efficiency E with 0.1–0.3 μm particles under rated flow rate, %	Initial pressure drop ΔP_i under rated flow rate, Pa	Note
D	99.999	≤ 250	Leakage scanning
E	99.9999	≤ 250	Leakage scanning
F	99.99999	≤ 250	Leakage scanning

3.4.3   Experimental Validation for Application of HEPA Filter with Circulation Air

1. Experimental method

   Experiment was performed in the aforementioned simulated isolation ward as shown in Fig. 3.17. Bacteria were released outside of the air filter for exhaust air. The bacterial concentration for spray was 8×10^8 pc/mL.

   Leakage may occur on the frame of the air filter as shown in Fig. 3.31, which may not be detected beforehand and then sealed (a leakage-free exhaust air apparatus was invented, which will be introduced in detail later). In order to prevent the spread of aerosol into the room from the bacterial release position at the exhaust (return) air outlet, casing pipe must be used for bacteria generation inside.

   If the frame of air filter is very air-tight, the casing pipe should contact air filter directly, which is shown in Fig. 3.32 [24]. In the figure, Q is the flow rate of the circulation air in the room, m³/h; q_1 is the air flowrate during the spray, m³/h; q_2 is the air flowrate sucked into the casing pipe, m³/h; k is the penetration of HEPA filter at the return air outlet, %; C is the bacterial generation rate, pc/h.

   It is obvious that the aerosol concentration in the supply air is kC/Q.

   When there is leakage air on the frame of air filter or when the casing pipe does not closely contact the air filter as shown in Fig. 3.33 [24], is there any influence on the measured result?

   In Fig. 3.33, q_1 is the air flowrate during the spray, m³/h; q_2 is the air flowrate sucked into the casing pipe, m³/h; q_3 is the flowrate of the leakage air, m³/h; Q is the flow rate of exhaust air through air filter, m³/h; k is the penetration of air filter, %; C is the bacterial generation rate, pc/h; Q/X is the flowrate inside the casing pipe ($Q/X = q_1 + q_2$, $Q = q_2 + Q/X$), where X is the ratio which is larger than 1.

Table 3.26 Approximated comparison of several standards for air filters from China and abroad

Chinese standard	EU standard Euvovent 4/9	ASHRAE standard with arrestance, %	ASHRAE standard with dust spot method, %	U.S. DOP method (0.3 μm), %	EU standard EN779	German standard DIN 24185	U.S. standard MERV
Coarse filter 4	EU1				G1	A	1
Coarse filter 3	EU1	<65			G1	A	2–4
Coarse filter 2	EU2	65–80			G2	B1	5–6
Coarse filter 1	EU3	80–90			G3	B2	7–8
Fine filter 3	EU4	≥90			G4	B2	9–10
Fine filter 2	EU5		40–60	20–55	G5	C1	11
Fine filter 1	EU6		60–80		F6	C1/C2	12
High and medium efficiency filter	EU7		80–90	55–60	F7	C2	13
High and medium efficiency filter	EU8		90–95	65–70	F8	C3	14
High and medium efficiency filter	EU9		≥95	75–80	F9	–	14
Sub-high efficiency filter	EU10			>85	F10	Q	15
Sub-high efficiency filter	EU11			>98	F11	R	16
HEPA filter A	EU12			>99.9	F12	R/S	17
HEPA filter A	EU13			>99.97	F13	S	17
HEPA filter B	EU14			>99.997	F14	S/T	18–19
HEPA filter C	EU15			>99.9997	U15	T	19
ULPA filter D	EU16			>99.99997	U16	U	–
ULPA filter E–F	EU17			>99.999997	U17	V	–

Fig. 3.31 Leakage on the frame of air filter for exhaust (return) air

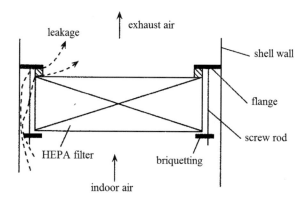

Fig. 3.32 Direct contact of casing pipe with air filter

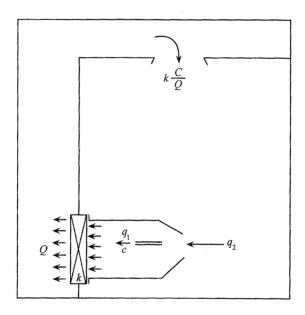

In this case, the concentration of entering air from return air opening is $C/(Q/X)$, pc/m^3. The penetrated aerosol in air is $kC/(Q/X)$. The total quantity of aerosol penetrated is:

$$\frac{kC}{\frac{Q}{X}} \cdot \frac{Q}{X} / Q = \frac{kC}{Q}$$

This means that the concentration is the same as that from the supply air for the case when the casing pipe is required to contact air filter directly and when there is no leakage on frame of air filter. This proves that it is feasible to use casing pipe for bacterial generation.

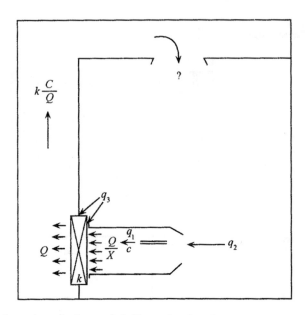

Fig. 3.33 Leakage air on the frame of air filter and casing pipe

With the same method, bacterial concentration can be measured below the air supply outlet with the sampling method for planktonic bacteria, which is shown in Fig. 3.34.

2. Experimental results

The measurement condition is shown in Table 3.27 [24]. The measurement results are presented in Table 3.28 [14]. In the table, the measured concentration for the supply air at the air supply outlet is expressed with CFU/petri dish.

Hourly bacterial generation rate at the return air opening = Bacterial solution concentration per minute × Spray quantity of bacterial solution × 30 min × 2

Bacterial generation concentration at the return air opening = Bacterial generation rate at the return air opening ÷ flow rate of return air

It is shown from Table 3.27 that for HEPA filter B, the filtration efficiency for the spray strain reached 99.99997%. For HEPA filter C, the filtration efficiency for the spray strain reached 99.999997%. Both efficiency of these two kinds of filters are larger than the measured efficiency with atmospheric dust. This is consistent with the aforementioned factor of equivalent diameter of microorganism. Filtration efficiency of HEPA filter C for bacteria is higher than that of HEPA filter B by one order of magnitude. This is also consistent with the relationship of the filtration efficiency with atmospheric dust between two products.

Next quantitative analysis will be performed.

The relationship between different particle diameters is given by empirical equation [25].

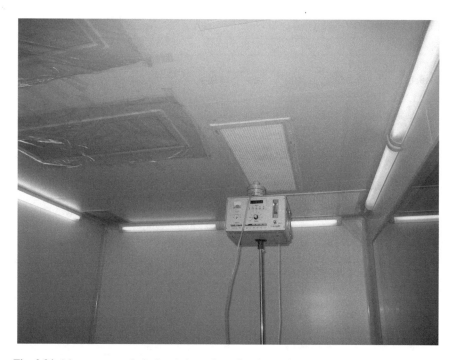

Fig. 3.34 Measurement of planktonic bacteria at the air supply outlet

$$k_2 \frac{k_1}{e^{(d/d_{0.3})^2}} \tag{3.9}$$

where k_1 and k_2 are the penetrations for particle size 0.3 μm and particle size larger than 0.3 μm, respectively; $d_{0.3}$ and d are the particle size 0.3 μm and the particle size larger than 0.3 μm, respectively;

From the above experiment shown in Table 3.27 [14], the measured efficiency of HEPA filter B before delivery from factory for atmospheric dust with diameter ≥ 0.5 μm is 99.999%. From literature [25], the efficiency with atmospheric dust for particle size 0.3 μm can be calculated to be 99.93%, which corresponds to the penetration $k = 0.07\%$. The measured efficiency of HEPA filter C before delivery from factory for atmospheric dust with diameter ≥ 0.5 μm is 99.99994%. The efficiency with atmospheric dust for particle size 0.3 μm can be calculated to be 99.998%, which corresponds to the penetration $k = 0.002\%$.

From Chap. 5, water component in the sprayed droplet will evaporate quickly during the spray process, but residual solute will be left. Since the size of the sprayed droplet is usually 1–5 μm, the average size is 3 μm. From Chap. 5, the final size of the solute is $0.16 \times 3 = 0.48$ μm, so it should be added into the size of naked bacteria. Since the size of the spore is about 1 μm, the increase of the linear thickness on the surface can be ignored.

Table 3.27 Experimental condition

No. of experiment	1	2	3
Type of HEPA filter	B	C	B
Particle counting efficiency of HEPA filter before delivery from factory (≥ 0.5 μm)	99.999	99.99994	99.999
Bacterial solution concentration (pc/mL)	8×10^{10}	4.5×10^{10}	4.5×10^{10}
Spray volume of bacterial solution (mL/min)	0.204	0.159	0.153
Air flowrate during bacterial generation (L/min)	17	17	17
Number of vessel in Andersen sampler	One vessel in each layer, 2 layers	One vessel in each layer, 2 layers	One vessel in each layer, 2 layers
Flow rate through Andersen sampler (L/min)	28.3	28.3	28.3
Number of vessel in centrifugal sampler (Type WL 1)	One vessel in each layer	One vessel in each layer	One vessel in each layer
Flow rate through centrifugal sampler (Type WL 1) (L/min)	28.3	28.3	28.3
Period of bacterial generation (min)	30	63	36
Volume of bacterial solution (mL)	6.12	10	5.5
Sampling position	15 cm from the center of the air supply outlet	15 cm from the center of the air supply outlet	15 cm from the center of the air supply outlet
Flow rate of return air (m^3/h)	380	223	233

Table 3.29 illustrates the efficiency for spores with different diameters.

Based on the measurement result from Table 3.28, the measured efficiency is within the range of the calculated efficiency for spores with diameter 0.8 and 0.9 μm. From Table 3.24, the concentration at the air supply outlet, when HEPA filter C was applied, was in the order of the natural number magnitude. This means for such high efficiency in experiment, when the bacteria concentration was increased by an order of magnitude, the measured efficiency may become higher.

The above measured result is equivalent with the foreign experimental data in Table 3.30 [26]. The corresponding filtration velocity is only less than a half of the filtration velocity in Table 3.30, so the efficiency value is slightly higher.

The above efficiency values with the sodium flame method, DOP method and particle counting method are different. For particle counting method, the efficiency for particle size 0.3 μm is different from that for particle size ≥ 0.5 μm, which could be referred to literature [25].

Table 3.28 Measurement result for filtration efficiency with bacteria

Type of air filter	Sampling method	Sampling time, min	Concentration at air supply outlet, CFU/dish or CFU/m³	Bacterial generation rate at return air opening, CFU/h	Bacterial concentration released at return air opening, CFU/m³	Filtration efficiency, %	Average filtration efficiency, %	No. of experiment
B	Andersen sampler	1	41 (41 × 1000/28.3 = 1448)	9.79×10^{11}	2.58×10^{9}	99.999944	99.999989	1
		3	91 (91 × 1000/28.3 × 3 = 1072)			99.999958	99.999949	
		5	200 (200 × 1000/28.3 × 5 = 1414)			99.999945	99.99997	
B	Andersen sampler	1	5.17 (5.17 × 1000/28.3 = 182)	4.14×10^{11}	1.8×10^{9}	99.99999	99.999991	3
		3	4.5(4.5 × 1000/28.3 × 3 = 53)			99.999992		
	Andersen sampler	1	4.7(4.7 × 1000/28.3 = 166)	4.29×10^{11}	1.9×10^{9}	99.999991	99.99999	
		3	17 (17 × 1000/28.3 × 3 = 200.2)			99.999989		
C	Andersen sampler	3	4.8 (4.8 × 1000/28.3 × 3 = 56.5)	4.29×10^{11}	1.9×10^{9}	99.999997	99.999997	2
	Centrifugal sampler	3	8(8 × 1000/28.3 × 3 = 94.2)			99.999995	99.999996	
	Centrifugal sampler	5	11.3 (11.3 × 1000/28.3 × 5 = 80.1)	4.29×10^{11}	1.9×10^{9}	99.999996		

Table 3.29 Efficiency for spores with different diameters

Air filter	Calculated efficiency, %				Measured efficiency, %
	0.5 μm	0.8 μm	0.9 μm	1.0 μm	
B	99.9957	99.999946	99.9999916	≈100	99.99997
C	99.99988	99.9999952	99.9999993	≈100	99.999997

Table 3.30 Filtration efficiency of various air filters for serratia marcescens (the bacterial solution concentration for spray is 1.1×10^7 pc/L)

Type of air filter	Times of experiments	Efficiency, %	Filtration velocity, m/s	Note
DOP 99.97 DOP 99.97	20 19	99.9999 99.9994 ± 0.0007	0.025 0.025	Equivalent to HEPA filter B in China
DOP 99.97 DOP 95	20 17	99.996 ± 0.0024 99.989 ± 0.0024	0.025 0.025	Equivalent to sub-high efficiency air filter in China
DOP 75 DOP 60 DOP 40	20 20 20	99.88 ± 0.0179 97.2 ± 0.291 83.8 ± 1.006	0.05 0.05 0.05	Equivalent to high-medium efficiency air filter in China
DOP 20–30	18	54.5 ± 4.903	0.20	Equivalent to or slightly better than fine filter in China

In short, the theoretical efficiency with actual particle size should be larger than the experimental data. So it is much safe to use experimental data.

It is known that the number of aerosol during coughing could reach more than 3×10^5. When ten times of this concentration, i.e., 3 million bacterial particles, pass through HEPA filter B, only one particle could penetrate. When one hundred times of this concentration, i.e., 30 million bacterial particles, pass through HEPA filter C, only one particle could penetrate.

Therefore, the bacterial concentration in the circulation air through HEPA filtration unit is much smaller than indoor bacterial concentration. For patients staying in a room with such low bacterial concentration for a long time, it is not doubt that the influence of circulation air is little. This means it is feasible to apply HEPA filtration unit in isolation ward.

References

1. Z. Xu, in *Design, Operation and GMP Accreditation on Pharmaceutical Factory*, 2nd edn. (Tongji University Press, 2011), pp. 42
2. Z. Xu, in *Design Principle of Isolation Ward* (Science Press, Beijing, 2006), pp. 5–18
3. Z. Xu, in *Fundamentals of Air Cleaning Technology*, 4th edn. (Science Press, Beijing, 2014) pp. 401
4. Z. Xu, in *Design, Operation and GMP Accreditation on Pharmaceutical Factory*, 1st edn. (China Architecture & Building Press, Beijing, 2002) pp. 208
5. E. Moia, Keypoints of biological isolation facilities, in *Proceedings of the 6th China International (Shanghai) Academic Forum & Expo on Cleanroom Technology*, 2003, p. 317
6. T. Zhao, K. Jia, Discussion on the pressure control of BSL-3 laboratory, Build. Sci. no. supplementary issue, pp. 112–115 (2005)
7. Z. Xu, Y. Zhang, Q. Wang, H. Liu, F. Wen, X. Feng, Isolation principle of isolation wards. *J. HV&AC*, **36**(1), pp. 1–7, 34 (2006)
8. Z. Xu, Y. Zhang, Q. Wang, F. Wen, H. Liu, L. Zhao, X. Feng, Y. Zhang, R. Wang, W. Niu, Y. Di, X. Yu, X. Yi, Y. Ou, W. Lu, Study on isolation effects of isolation wards (1). J HV&AC **36**(3), 1–9 (2006)
9. Z. Xu, The application of surge chamber of negative pressure isolating room. Chin. Hosp. **10**(10), 17–20 (2006)
10. Z. Xu, in *Fundamentals of Air Cleaning Technology and Its Application in Cleanrooms.* (Springer Press, 2014), pp. 449–454
11. J.A. Schmidt, in *System and Method of Applying Energetic Ions for Sterilization*, Official Gazette of the United States Patent & Trademark Office Patents (2003)
12. Z. Xu, Y. Zhang, Q. Wang, H. Liu, F. Wen, X. Feng, Y. Zhang, L. Zhao, R. Wang, W. Niu, D. Yao, X. Yu, X. Yi, Y. Ou, W. Lu, Study on isolation effects of isolation wards (3). J. HV&AC **36**(5), 1–4 (2006)
13. L. Zhao, Z. Xu, X. Yu, Q. Wang, Y. Zhang, F. Wen, H. Liu, Y. Di, R. Wang, H. Zhao, Microbiology experimental method for insulation effect of isolation wards. J. HV&AC **37**(1), 9–13 (2007)
14. Y. Zhang, Z. Xu, Q. Wang, H. Liu, F. Wen, L. Zhao, X. Feng, Y. Zhang, R. Wang, W. Niu, Y. Di, H. Zhao, X. Yu, X. Yi, Y. Ou, W. Lu, Experiment on germ filtering efficiency of high efficiency filters on return air inlet in isolation wards. *J. HV&AC*, **36**(8), pp. 95–96 + 112 (2006)
15. S. Honda, Y. Kita, K. Isono, K. Kashiwase, Y Morikawa, Dynamic characteristics of the door opening and closing operation and transfer of airborne particles in a cleanroom at solid tablet manufacturing factories. *Trans. Soc. Heating, Air-Conditioning Sanit. Eng. Japan*, **95**, pp. 63–71 (2004)
16. Z. Xu, Y. Zhang, Y. Zhang, Z. Mei, J. Shen, D. Guo, P. Jiang, H. Liu, Mechanism and performance of an air distribution pattern in clean spaces. J. HV&AC **30**(3), 1–7 (2000)
17. H. Yang, "Implication of electrostatic sterilization in air cleaning," in *the 37th Pharmaceutical Preparations Forum & the 4th Pharmaceutical Disinfection and Sterilization Symposium*, 2009, pp. 127–128
18. C. Wu, M. Du, Effects of four kinds of indoor air sterilization method. Chin. J. Nosocomiology **10**(6), 403 (2000)
19. Y. Wang, in *Study on Air Sterilization Technique with Dynamic UV Irradiation in Air Duct.* (Tianjin University, 2011)
20. H. Pan, Discussion on applicability of sustained using ultraviolet rays in air conditioned rooms for disinfecting and sterilization: part 6 of the series of research practice of the revision task group of the architectural technical code for hospital clean opera. *J. HV&AC*, **43**(7), pp. 27–29 + 36 (2013)

21. Z. Xu, *Fundamentals of Air Cleaning Technology*, 4th edn. (Science Press, Beijing, 2014)
22. H. Mao, in *Study on Improvement of Indoor Air Quality by Application of Electrostatic Air Cleaner*. (Tongji University, 2008)
23. J. Mao, J. Shen, Control concept and practice for the sterilized space. Contam. Control Air-Conditioning Technol. **4**, 9–12 (2003)
24. Z. Xu, *Design Principle of Isolation Ward*. (Science Press, Beijing, 2006), pp. 149
25. Z. Xu, in *Fundamentals of Air Cleaning Technology and Its Application in Cleanrooms*. (Springer Press, 2014, 2014), pp. 205–207
26. Z. Xu, *Fundamentals of Air Cleaning Technology and Its Application in Cleanrooms*. (Springer Press, 2014), pp. 499. Fig. 3.1 Schematic diagram of gap
27. J. Shen, Multi-application isolation ward and its air conditioning technique without condensed water. Build. Energy Environ. **24**(3), 22–26 (2005)

Chapter 4
Air Distribution Design in Negative Pressure Isolation Ward

4.1 Fundamental Principle of Air Distribution in Negative Pressure Isolation Ward

The volume of the isolation ward is not large, so the air distribution is relative simple. But the fundamental principle to control air pollution should be followed at first. The total trendy of air flow is consistent with the direction of gravitational deposition on pollutant, so that pollutant can be removed in the fastest way and in the shortest distance.

The size of the droplet nuclei aerosol generated by patient in the isolation ward is usually between 0.001 and 100 μm. Solid, liquid or the combined solid/liquid aerosol with the density $\rho = 1 \sim 2$ g/cm^3 are much heavier than the air molecular. The transmission property of these aerosol is much likely.

Therefore, the air distribution schemes with up-supply and down-return (or exhaust), ceiling air supply and down-side wall air return (or exhaust), upside-side air supply and floor return (or exhaust) are usually adopted in the isolation ward.

It was recommended in the ASHRAE manual in 1991 that air supply outlet should be installed in the ceiling so that air can be supplied towards the sensitive area, the ultra-clean region and the seriously polluted region. Meanwhile, the surrounding exhaust air opening or the single exhaust outlet should be set near the floor. In this way, clean air could be supplied to the breathing zone and the working area, then downwards to the polluted floor area to be exhausted.

It was also pointed out by scholar Chan Fan during the research after SARS epidemic that the system with up-supply and down-return could effectively dilute and remove pollutant [1]. Simulation result with computational fluid dynamics showed that this kind of design could reduce the crossflow of the polluted air in the SARS ward. It was also found that in the breathing area of the medical personnel at the position 1.5 m above the floor, the particle concentration containing virus was greatly reduced.

© Springer Nature Singapore Pte Ltd. 2017
Z. Xu and B. Zhou, *Dynamic Isolation Technologies in Negative Pressure Isolation Wards*, DOI 10.1007/978-981-10-2923-3_4

The air distribution scheme with down-supply and up-return, which is usually used in the computer room, has never been adopted yet in air cleaning and pollution control techniques.

The air distribution scheme with upper-supply and upper-return is used in the corridor and applications where occupant activity is rare without generation of pollutant. But in air cleaning and pollution control technique, it is usually not recommended. From the above examples, at least the following aspects of short-comings can be found.

(1) The number of large particles with diameter 5 μm is relative large at the position with certain height (such as the breathing zone).
(2) Air velocity in the working area is usually small.
(3) The time for self-purification is long. Field test showed that it could be pro-longed by one time.
(4) Short circuit of flow will occur, so that part of the supplied air and the outdoor air have no effect in the room.
(5) During the rising process, pollutant particles will cause pollution at places where they pass through.

This has also been noticed in "*Guidelines for design and construction of hospital and health care facilities*" by AIA in U.S.A. that when air supply outlet and air exhaust opening are placed on the ceiling or on the opposite side walls, short circuit will appear. In this way, the ventilation effect and the cooling/heating capacity are reduced in essence. The air distribution in the habitant room of the patient is of the same significance as the flow rate supplying into the room.

Next the detailed analysis for setting of the air supply and air exhaust openings will be presented in theory.

4.2 Velocity Field Near Return Air Opening

The key to the aforementioned problems is the position of the return air opening. Therefore, the basic understanding for the velocity field near the return air opening should be established.

Here the common analytic method is used to analyze the flow field near the return air opening.

Return air opening is the limited sink. At first, according to the fluid dynamics, when the size of the suction inlet is very small, it could be considered as "point sink". When there is no resistance for the flow near the suction inlet, the working surface of the suction inlet is spherical. The air velocity V_x at the distance x (m) from the suction inlet is

$$V_x = \frac{Q}{4\pi x^2} \tag{4.1}$$

when V_0 is the average air velocity at the suction inlet and F_0 is the area of the suction inlet, we obtain

$$\frac{V_x}{V_0} = \frac{F_0}{4\pi x^2} \tag{4.2}$$

When the suction inlet is the return (or exhaust) air opening indoors, one side will be limited. The flow field is a half of the free sucked flow for the area $2F_0$ [2]. In the above equation, F_0 is replaced by $2F_0$.

Meanwhile, the flow will result in the resistance. Since the opening has a certain size, the actual iso-velocity surface will become ellipsoidal from spherical, which is shown in Fig. 4.1.

Experiment has shown that the calculated flow field with the above equation is close to the measured value [2]. When $x/d_0 > 0.5$, Eq. (4.1) can be used to calculate the air velocity at each point. For circular air supply opening, d_0 is the diameter of air supply outlet. For rectangular air supply opening, d_0 is the length of the short side. When $x/d_0 > 1.5$, the actual attenuation rate of air velocity is larger than the calculation result [3].

Suppose the area of the room is 15 m² and the height is 2.6 m. There is one door and one delivery window in the room. The flow rate of the exhaust air from the room is only 120 m³/h [3]. The size of two return air grilles is 2 m × 0.4 m × 0.4 m. Their shading coefficient is 0.9. The average air velocity of each return air grille is $V_0 = 0.12$ m/s. The corresponding air velocity is 0.2 m/s when the flow rate is 200 m³/h.

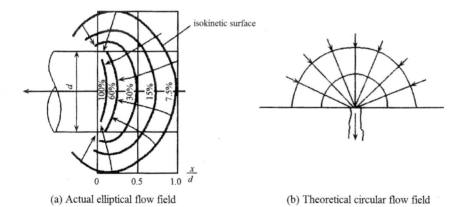

(a) Actual elliptical flow field   (b) Theoretical circular flow field

Fig. 4.1 Flow field near the return air opening

When the height of the breathing zone is 1.4 m, we obtained

$$\frac{x}{d_0} = \frac{2.6 - 1.4}{0.4} = 3$$

So we obtain

$$V_x = V_0 \frac{2F_0}{4\pi x^2} = \frac{(0.12 \to 0.2) \times 2 \times 0.4^2}{4\pi (3 \times 0.4)^2} = 0.002 \to 0.0033 \, \text{m/s}$$
$$= 0.2 \to 0.33 \, \text{cm/s}$$

The air velocity in the breathing zone because of the rising flow by the return air grille becomes from 0.2 to 0.33 cm/s. As mentioned before, the actual attenuation rate of air velocity may be larger than the calculated value. So the air velocity of the rising flow is less than 0.2–0.33 cm/s.

4.3 Velocity Decay Near Air Supply Outlet

Compared with the attenuation rate of air velocity near the return air opening, the attenuation rate of air velocity near the air supply outlet is much small. According to the calculation by Zhang Yan-guo, the simulated air velocities in a laboratory with area 8 m^2 are shown in Figs. 4.2, 4.3, 4.4, 4.5, 4.6, 4.7, 4.8, 4.9, 4.10, 4.11, 4.12, 4.13, 4.14, 4.15, 4.16, 4.17, 4.18, 4.19 and 4.20 [4, 5]. In these figures, V_a is the air velocity in the working area with height 0.8 m.

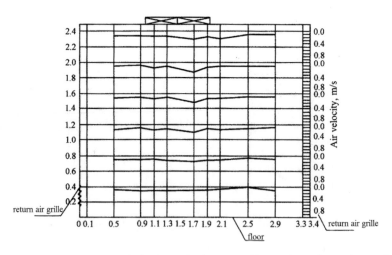

Fig. 4.2 Measured attenuation of air velocity for the room with two air supply openings and with air cleanliness level 100,000

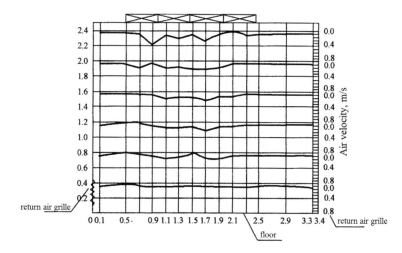

Fig. 4.3 Measured attenuation of air velocity for the room with four air supply openings and with air cleanliness level 10,000

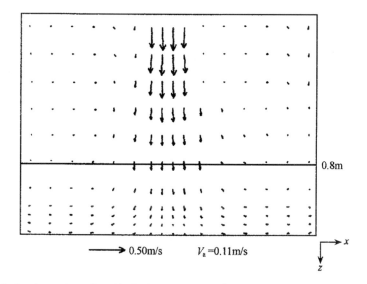

Fig. 4.4 An air supply outlet with air change rate 10 h^{-1}

From these figures, for the common number of air supply outlets, the air velocity will decay to be $V_a = 0.1$–0.2 m/s after the supplied air reached the height 0.8 m (which is usually called the working area).

The attenuation relationship between the distance from the air supply opening and the air velocity can be found with the equation given by Liu [6], i.e.,

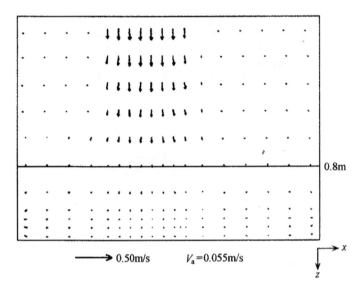

Fig. 4.5 Two air supply outlets with air change rate 10 h^{-1}

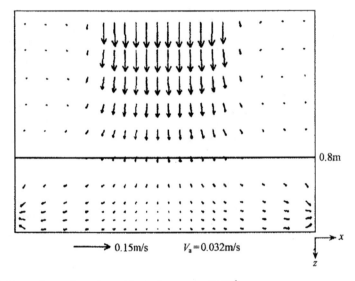

Fig. 4.6 Three air supply outlets with air change rate 10 h^{-1}

$$\frac{V}{V_0} = 1 - 0.09\frac{x}{r} \tag{4.3}$$

where x is the distance from the center of the air supply outlet and the wall; r is the equivalent radius of the air supply outlet.

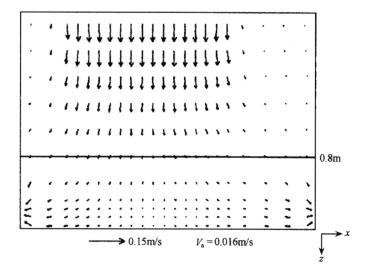

Fig. 4.7 Four air supply outlets with air change rate 10 h^{-1}

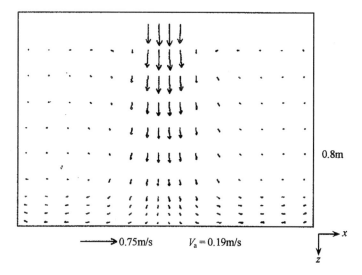

Fig. 4.8 One air supply outlet with air change rate 15 h^{-1}

For the common air supply outlet (which has smaller area than the case for local air cleanliness of one hundred, so the value of r is also small), the measured attenuation rate of air velocity is shown in Table 4.1.

In short, the air velocity in the working area with upper-supply and down-return scheme is more than that with upper-supply and upper-return (or exhaust) scheme by hundreds times.

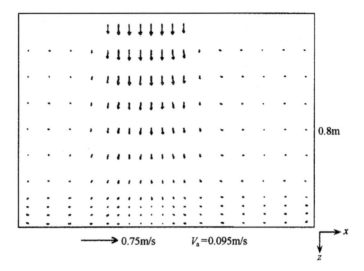

Fig. 4.9 Two air supply outlets with air change rate 15 h^{-1}

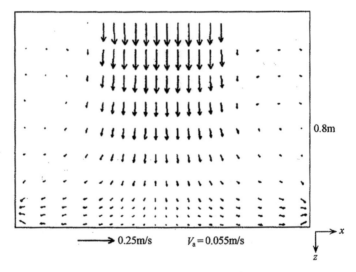

Fig. 4.10 Three air supply outlets with air change rate 15 h^{-1}

4.4   The Following Speed and the Deposition Velocity

Under the action with three kinds of forces including the gravitational, inertial (mechanical) and diffusional forces, the velocity of the particle itself and the distance of movement are very small. For a particle with diameter 1 μm, the distances of movement per second are about 0.006, 0.0006 and 0.0004 cm. But the air

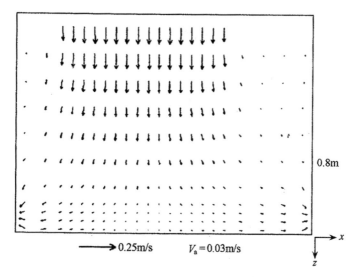

Fig. 4.11 Four air supply outlets with air change rate 15 h^{-1}

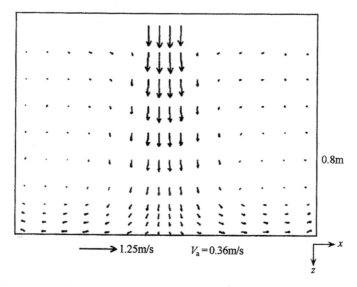

Fig. 4.12 One air supply outlet with air change rate 25 h^{-1}

velocity of indoor air (including the convective velocity of thermal flow) is usually more than 0.1 m/s. In the air flow, small particles are almost transmitted with the same velocity as the velocity of the air flow.

According to theoretical calculation, for particles with diameter 5 μm when the density ρ is 1 g/cm^3, the following speed is equivalent to 90% of the velocity of air

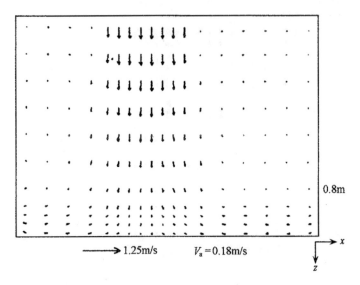

Fig. 4.13 Two air supply outlets with air change rate 25 h^{-1}

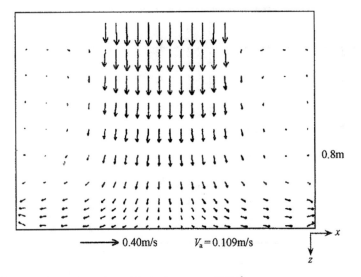

Fig. 4.14 Three air supply outlets with air change rate 25 h^{-1}

flow. For particles with diameter 1 μm, the following speed reaches 99.9% of the velocity of air flow [7]. Even when the density ρ is more than twice, the following speed reaches 90% of the former case. So for particles with diameter 5 μm with this density value, the following speed is still 80% of the air flow velocity.

Therefore, for many particles with diameter less than 5 μm, when air flows upwards (in upper-return or exhaust situation), particles will be exhausted upwards

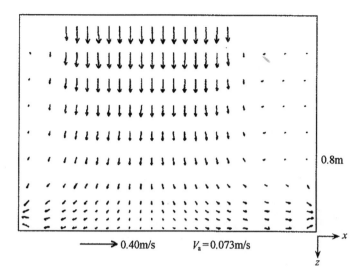

Fig. 4.15   Four air supply outlets with air change rate 25 h^{-1}

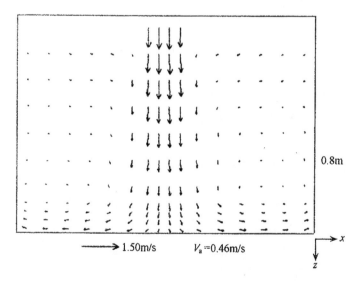

Fig. 4.16   One air supply outlet with air change rate 30 h^{-1}

too. The difference between the trajectory of particles and the streamlines is very small, which is shown in Fig. 4.21 and Table 4.2.

For atmospheric dust with density $\rho = 2$ g/cm^3, the deposition velocity V_s of particles can be calculated as follows [8].

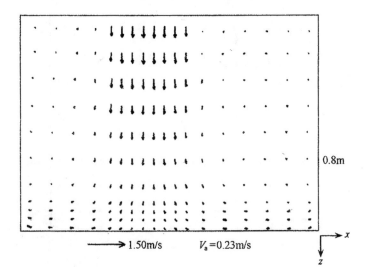

Fig. 4.17 Two air supply outlets with air change rate 30 h^{-1}

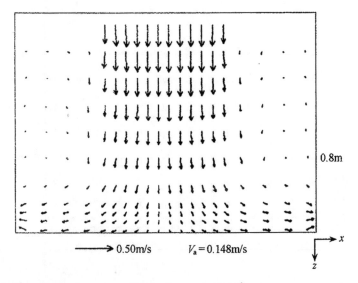

Fig. 4.18 Three air supply outlets with air change rate 30 h^{-1}

$$V_s = 0.6 \times 10^{-2} \times d_p^2 \ (\text{cm/s})$$

where d_p is the particle diameter. When $d_p = 5 \ \mu m$, $V_s = 0.15$ cm/s.

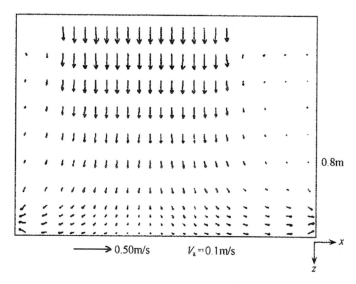

Fig. 4.19 Four air supply outlets with air change rate 30 h^{-1}

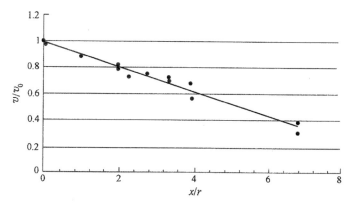

Fig. 4.20 Comparison between experimental data and fitted result for attenuation rate of air velocity

It is shown that the upward rising velocity 0.2 cm/s at the working height 0.8 m is equivalent to the deposition velocity 0.15 cm/s. So the phenomenon for self-locking of particles is understandable.

Table 4.1 Attenuation rate of air velocity in experiment

Air change rate, h^{-1}	Number of air supply openings	Designed air velocity at air supply opening, m/s	Ratio of air supply area	Air velocity in the mainstream area with height 0.8 m, m/s	Attenuation rate of air velocity	Average attenuation rate	
						Experiment	Simulation
15	1	0.54	0.03	0.32	0.41	0.43	0.48
	2	0.27	0.06	0.15	0.44		
25	1	1.02	0.03	0.55	0.46	0.45	0.44
	2	0.50	0.06	0.28	0.44		
	3	0.34	0.09	0.02	0.41		
	4	0.25	0.12	0.13	0.48		

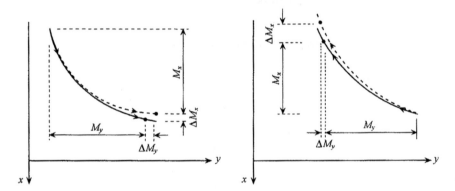

Fig. 4.21 Schematic of the following speed

Table 4.2 Variation of the trajectory of particles and the streamlines

Particle diameter		For air velocity 0.3 m/s
1.004 mm	$\frac{\Delta M_x}{M_x}$	35.6×10^{-5}
	$\frac{\Delta M_y}{M_y}$	7.7×10^{-5}

4.5   Composition of Velocities and Vortex [9]

When particles are below the return (or exhaust) air opening, they will be influenced by the suction force by the return air opening, which is the carrying force by the upward air velocity. Meanwhile, they will also be affected by the gravitational force and the combined effect of the supplied air and the vortex. When the working area is not just below the return (or exhaust) air opening and it is located in one side of the room, the influence of vortex will be much larger.

When the force by supplied air is equivalent to that by return air, particles will have nearly horizontal air velocity, which is shown in Fig. 4.22. When the flow rate of supply air is larger than that of exhaust air and the air velocity in the working area reaches the magnitude of decimeter per minute, the direction of velocity on particles must be downward, which is shown in Fig. 4.23.

Therefore, for particles with diameter less than 5 μm, although the deposition velocity is small, with the influence of air velocity only horizontal movement or rotating movement will appear. Upward rising movement will never occur. This means that these particles will wander around the breathing zone of occupant.

Fig. 4.22 One schematic of force by airflow on particles

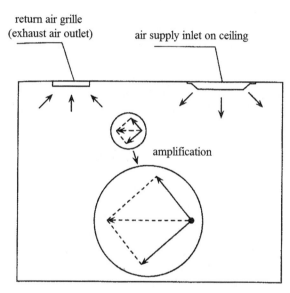

Fig. 4.23 Another schematic of force by airflow on particles

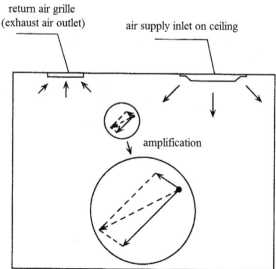

4.6   Position of Air Supply, Exhaust and Return Outlets in Isolation Ward

4.6.1   Fundamental Principle

It has been mentioned in the previous section that the fundamentals of air distribution in the isolation ward are that the total trend of air flow should be consistent with the deposition direction of pollutant by gravitational force. Therefore, the priority is the scheme with upper-supply and down-return.

With this general principle, there are many specific combination schemes for the positions of air supply and return (or exhaust) openings. This is the content which will be introduced in this section. First of all, the basic principle for this problem will be discussed.

The first basic principle is that unidirectional flow is easily realized with the designed positions of air supply and return (or exhaust) openings.

The concepts of directional flow and unidirectional flow in the air cleaning technology are different, which were confused before. The core of unidirectional flow is that the direction of flow is single, the streamlines are comparatively parallel, and the air velocities are relatively uniform. While the implication of unidirectional flow is that the total trend of airflow direction is fixed, which passes from the clean area towards the polluted area or through the clean area → potentially polluted area → polluted area. Streamlines are not required to be parallel, and air velocities are not required to be uniform. Figure 4.24 shows the unidirectional flow in the cleanroom. Figure 4.25 shows the directional flow in the cleanroom.

It is shown that in the flow field of directional flow, local vortex could exist. But the total trend of airflow is fixed, which passes from the clean area (near the position below the air supply outlet) to the polluted area (near the operation table).

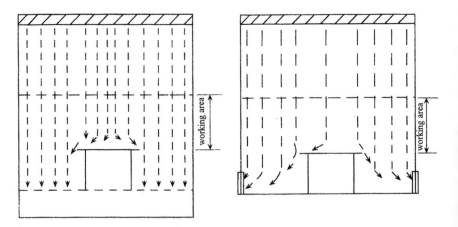

Fig. 4.24 Schematic of unidirectional flow

Fig. 4.25 Schematic of
directional flow

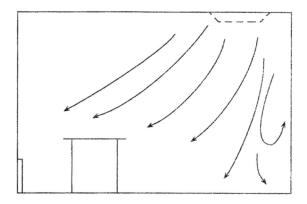

In the infectious isolation ward for microorganism pollution control, the CDC manual in U.S.A. published in 1994 clearly stated for the required directional flow that the ventilation system should be designed and balanced to force air passed through less polluted area (or relative clean area) to seriously polluted area (or relative unclean area). For example, air should flow from the corridor (relative clean area) to the isolation ward for pulmonary tuberculosis (unclean area), so that pollutant can be prevented from dispersing into other region. By setting relative low pressure in the area where air is supposed to flow into, the direction of airflow can be controlled.

Therefore, with the principle of directional flow, the scheme with exhaust air on single side wall is usually adopted.

Meanwhile, the manual from CDC provided specification on the positions of air supply and exhaust (or return) air openings based on the requirement of directional flow. One type of air distributions is to place the air supply opening at one side of the room, which is in the opposite direction facing the patient. Then air is exhausted at one side of the patient. Another kind of methods is that it is more efficient to supply air at lower temperature than that of the indoor air. In this case, air is supplied near the ceiling, and then it is exhausted near the floor.

Obviously the former method is the side air-supply and side air-exhaust scheme, and the latter is the top air-supply and down air-exhaust scheme.

For the side air-supply and side air-exhaust scheme (please refer to the figure later), ASHRAE manual has different opinions. It is believed that when air supply and exhaust openings are placed on the opposite side walls, short-circuit will occur from the supplied air to the exhausted air. This will reduce the ventilation performance and the cooling/heating capacity essentially. It can also be predicted that for upper air-supply and upper air-exhaust (or return), this shortcoming will be more obvious.

The second basic principle is that for more than one patient, they are not situated in upstream or downstream side of the airflow with the designed positions of air supply and return (or exhaust) openings. This is one important principle for prevention of cross-infection. For example, when two sickbeds are placed in parallel, and when air supply opening is placed at the left of the left sickbed while air exhaust opening is set at the right of the right sickbed, the right bed is located in the downstream of the left sickbed, which is prohibited.

The third basic principle is that it is beneficial to protect the medical personnel with the designed positions of air supply and return (or exhaust) openings, especially the potion of the air supply outlet. This has already been emphasized in the previous chapter.

As for as the air distribution is concerned, the CDC manual pointed out that in order to provide the optimal air distribution, the positions of the air supply and exhaust openings should be designed in such a way that clean air should flow through the possible working area of the medical personnel at first, and then flow through the pollution source, and enter into the exhaust air opening. In this case, the medical personnel will stay in the upstream of the airflow, although it is not possible to realize this kind of arrangement.

In the early version of ASHRAE manual in 1991, the necessity of directional flow with air supply from the breathing zone of the medical personnel at first has been pointed out. In usual condition, air is supplied from the air supply outlet installed on the ceiling, and then supplied to the sensitive ultra-clean area and the seriously polluted area, and exhausted by air exhaust opening near the floor. In this way, clean air flows through the breathing zone and the working area, and then flows downwards to the polluted flow area to be exhausted.

The forth basic principle is that the performance of the air distribution should be good enough for dilution purpose, and the draft sensation on patient should be avoided.

Three kinds of air opening positions were given in CDC manual (1994), which are shown in Figs. 4.26, 4.27 and 4.28. They are usually cited or imitated by researchers. It is required that the air exhaust (or return) openings are placed at the side wall near the head of patient (on the lower wall or middle of the wall). Meanwhile, the air supply openings are usually placed above the sickbed, or above the tail position of the sickbed, or on the wall near the tail of the sickbed. However, the issue of protecting the medical personnel is not considered in these schemes (Fig. 4.28).

Fig. 4.26 One form of the air distribution recommended by CDC in U.S.A

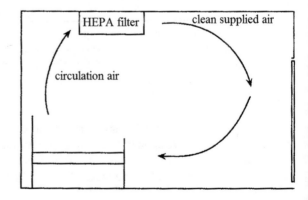

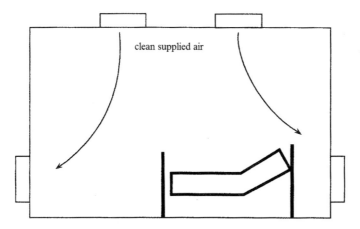

Fig. 4.27 The second form of the air distribution recommended by CDC in U.S.A

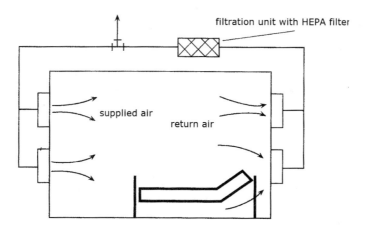

Fig. 4.28 The third form of the air distribution recommended by CDC in U.S.A

Moreover, experiment was performed by Jinming Shen and Weipeng Deng for the single side top-supply and opposite side lower-exhaust scheme, which is shown in Fig. 4.29 [10].

The theoretical analysis and numerical results on these schemes of air distribution by Deng [10] are summarized and presented in Table 4.3.

But among these schemes, how to prevent the unreasonable phenomenon shown in Fig. 3.28 is not considered.

It is firstly proposed the concept of protecting the medical personnel by the mainstream area in the research project on the isolation performance of the isolation ward [11].

Fig. 4.29 The air distribution with the single side top-supply and opposite side lower-exhaust scheme

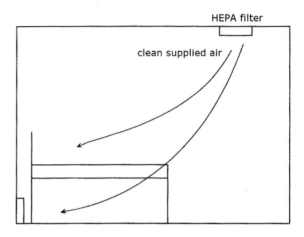

Table 4.3 Analysis on different air distributions

Scheme	Analysis	Simulated result
One form of the air distribution recommended by CDC in U.S.A.	Upward rising airflow is formed in the upper region of the room, which drves the exhaled droplet from the patient upward to the upper region above the sickbed. The ventilation efficiency of the room is the worst	
The second form of the air distribution recommended by CDC in U.S.A.	Strong mixed airflow is formed in the upper region of the room, which drives the exhaled droplet from the patient to disperse indoors. The ventilation efficiency of the room is worse. Draft sensation on patient face will appear	
The third form of the air distribution recommended by CDC in U.S.A.	The ventilation efficiency can be improved when the air velocity at the air exhaust opening above the sickbed is 1–2 m/s. But in this case the air velocity on the face of the patient reaches 0.5 m/s, which will cause the strong draft sensation	
The air distribution with the single side top-supply and opposite side lower-exhaust scheme	With optimal design, the performance of the double shutter is the best. The angle between the upper louver and the horizontal plane is 60°, while that between the lower louver and the horizontal plane is 40°	

4.6.2   Related Assessment Index

There are several assessment index for the performance of the combined positions of air supply and exhaust openings.

1. The dimensionless height when the pollutant concentration above the patient mouth reduces to 1% of the concentration in the breathing zone of patient [4].

The larger the height is, the easier the indoor pollutant disperses into the breathing zone of the medical personnel, and the larger the risk on the medical personnel standing by is.

$$\text{Dimensionless height} = \frac{\text{Height at the place with concentration equal to that near the mouth}}{\text{Room height}}$$

$$(4.4)$$

2. The dimensionless average concentration indoors at certain height

This dimensionless concentration reflects the dispersion and dilution velocity of the exhaled pollutant from patient, as well as the influence of air distribution.

$$\text{Dimensionless concentration} = \frac{\text{Indoor average concentration at certain height}}{\text{Concentration at breathing zone of patient}}$$

$$(4.5)$$

3. The ventilation efficiency in whole room

This value reflects the ability for removal of indoor pollutant by airflow. The larger the value is, the better the removal efficiency is. In this case, the quantity of the dispersed pollutant indoors is less. The ventilation efficiency value could be larger than 1.

$$E = \frac{N_E - N_S}{N_V - N_S}$$

where E is the ventilation efficiency in whole room; N_E is the pollutant concentration at the air exhaust (or return) opening; N_S is the pollutant concentration at the air supply opening; N_V is the average pollutant concentration indoors.

4. Predicted percentage of dissatisfied due to draft PD [10]

The value of PD is based on the draft sensation on patient face. It reflects the decay rate of air velocity and the sensation for uniformity of airflow.

$$PD = (34 - t)(v - 0.05)^{0.62}(3.14 + 0.37vT_u), \ \% \qquad (4.7)$$

where PD is the percentage of dissatisfied due to draft in a position inside the room, %; t is the temperature at this place, °C; v is the air velocity at this place, m/s; T_u is the turbulence intensity at this place, %. Usually T_u can be set 5%.

For comfort air-conditioning system, the air velocity near the face of people should be less than 0.12 m/s. It is pointed out in ISO7730 that the value of PD should not exceed 15%, so that the draft sensation can be avoided.

4.6.3 Results from Numerical Simulation [10, 12]

1. Geometric model of the isolation ward with single sickbed

The actual geometry of the isolation ward with single sickbed is 4.5 m × 3.15 m × 2.7 m. There is one sickbed inside (1.9 m × 0.9 m). The air supply outlet is placed on the ceiling (1 m × 0.32 m). The air return opening is set near the sickbed (0.5 m × 0.2 m). Given the actual condition occurred in the ward, the patient model was placed on the sickbed (During experiment, the model can stay with recumbent position. Aerosol can be exhaled from the nasal cavity), at each side of the sickbed one medical personnel model was placed.

Feng et al. proposed four schemes for positioning of air openings based on the principle of the mainstream area, and performed numerical simulation [12]. The relative positions of objects in these schemes are presented in Figs. 4.30, 4.31, 4.32, 4.33 and 4.34. The detailed information of these schemes are:

Scheme No. 1: Two air return openings. Air supply opening B, air return openings a and b, patient 2 and virtual medical personnel 1.
Scheme No. 2-a: One air return opening. Air supply opening A and air return opening b.
Scheme No. 2-b: Two air return openings. Air supply opening A and air return opening a and b.
Scheme No. 3: One air return opening. Air supply opening C and air return opening b.
Scheme No. 4: One air return opening. Air supply opening A and B, and air return opening b.

2. Simulated results for single sickbed

(1) Simulation condition

The air change rate was 12 h^{-1}. The size of the mouth was 0.02 m × 0.02 m. The respiratory rate of patient was 8 L/min. The temperature on the surface of patient model was 37 °C. It is an unfavorable condition with the temperature 37 °C.

Fig. 4.30 Schematic of scheme No. 1 with two return air openings. (Air supply opening B was placed above the tail position of the sickbed. Rectangular air return openings a and b were set at both side near the head of the sickbed)

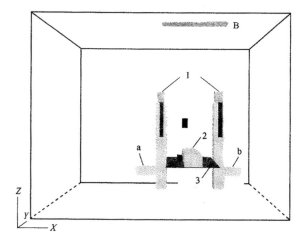

Fig. 4.31 Schematic of scheme No. 2-a with one return air opening. (Rectangular air supply opening A was placed above the standing position of the medical personnel. The position of air return opening b was the same as Scheme No. 1)

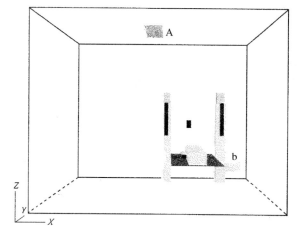

It was the surface temperature during the quilt is uplifted for health examination when the patient caught the fever with high body temperature. If the temperatures on the surface of patient model were set the same as the air, the simulated result will be unrealistic.

(2) Dimensionless height

Comprehensive comparison of dimensionless heights in various schemes is illustrated in Fig. 4.35, where the value of h is based on the position of patient mouth.

From the figure, the dimensionless height for scheme No. 2-a with one return air opening is the smallest, which is the most favorable for protection of medical personnel.

Fig. 4.32 Schematic of scheme No. 2-b with two return air openings. (The position of air supply opening A was the same position in the above scheme. Rectangular air return openings *a* and *b* were set at positions as Scheme No. 1)

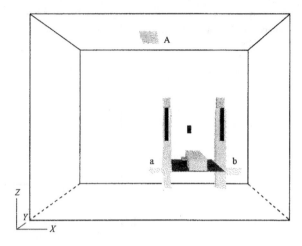

Fig. 4.33 Schematic of scheme No. 3 with one return air opening. (Air supply opening *C* was placed on the ceiling above the door. Air return opening *b* was set the same place as Scheme No. 1)

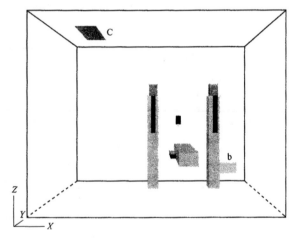

(3)   Dimensionless concentration

It is the ratio of the average pollutant concentration indoors to that near the mouth of patient at certain height for various schemes, which is shown in Table 4.4. In the table, the height is based on the floor.

From Table 4.4, the dimensionless concentrations at certain height for scheme No. 2-a with one air return opening and No. 4 are the lowest.

It was found in simulation that for scheme No. 2-a with one air return opening, the area where dimensionless concentration was larger than 10% is the smallest, and the corresponding dimensionless height in the vertical direction is the lowest. But for scheme No. 2-bN with two air return openings, short-circuit appears for part of the airflow, which worsens the performance of the air distribution indoors and elevates the height for dispersion of pollutant by 16.8%.

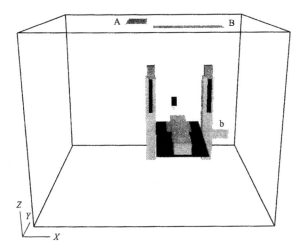

Fig. 4.34 Schematic of scheme No. 4 with one return air opening. (Air supply opening A and B were placed in the same places as Scheme No. 1 and No. 2. Air return opening *b* was set the same place as Scheme No. 1)

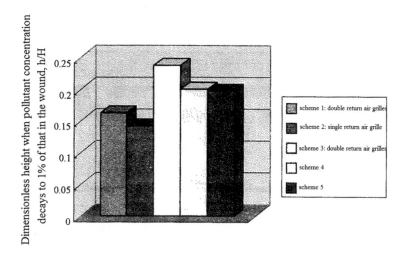

Fig. 4.35 Comparison of dimensionless height when pollutant concentration decays to 1 % of that in the wound

(4)   Situation with two sickbeds

Simulated results show that for two sickbeds, under the condition of normal respiratory rate (8 L/min), the sickbed A has no influence on the sickbed B. When the respiratory rate exceeds 20 L/min, the heights of pollutant dispersion near the patient for all the schemes are larger than 1.5 m, which is beyond the breathing zone of the standing occupant. In this case, risk will be formed for the medical personnel.

Table 4.4 Average dimensionless concentration indoors at certain height for various schemes

Height	0.65 m	1.1 m	1.5 m	$\sum$
Scheme No. 1: two air return openings	4.11×10^{-2}	2.33×10^{-2}	5.53×10^{-3}	0.000699
Scheme No. 2-a: one air return opening	3.58×10^{-2}	3.13×10^{-3}	8.15×10^{-5}	0.000399
Scheme No. 2-b: two air return openings	5.12×10^{-2}	5.86×10^{-2}	6.67×10^{-4}	0.0011
Scheme No. 3: one air return opening	3.50×10^{-2}	3.61×10^{-2}	1.66×10^{-2}	0.000877
Scheme No. 4: one air return opening	2.59×10^{-2}	3.39×10^{-2}	9.68×10^{-6}	0.000597

When the respiratory rate exceeds 80 L/min, mutual influence will appear for exhaled pollutant in various schemes.

Therefore, under most conditions, the cross infection between multi-bed can be avoided.

(5) Ventilation efficiency in the whole room

Based on the simulated result by Deng [10], the ventilation efficiencies with six different positions of air supply and return openings are shown in Table 4.5.

From Table 4.5, the ventilation efficiency with the up-supply and up-return scheme is the lowest, which is consistent with the conclusion in Sect. 4.1.

(6) Perceived percentage of dissatisfied due to local draft

The perceived percentage of dissatisfied was obtained for six different positions of air openings by Deng [10], which is shown in Table 4.6.

From Table 4.6, when the perceived percentage of dissatisfied is required to be less than 5%, the air velocity near the face should not exceed 0.1 m/s.

According to the research by Shen, Deng and Tang [10, 13, 14], the ventilation performance with displacement effect for the double shutter is much better than that of other air openings. A certain angle of supplied air can be formed with the double shutter, which could reduce the draft sensation on the surface of patient (refer to the next section). When the angle between the upper louver and the horizontal plane is 60° and when the angle between the lower louver and the horizontal plane is 40°,

Table 4.5 Comparison of the ventilation efficiency for six different positions of air supply and return openings in the isolation ward

Ventilation efficiency in the whole room	The single side top-supply and opposite side lower-exhaust scheme				Up-supply and up-return with high efficiency air cleaner (CDC from U.S.A.)	Up-supply and two-side return (CDC from U.S.A.)
	Single deflection grille	Double shutter	Square ceiling diffuser	Low-velocity perforated ceiling supply air outlet		
E	1.17	2.06	0.85	0.71	0.29	0.50

Table 4.6 Comparison of PD on patients with six different positions of air openings in isolation ward

Assessment index	The single side top-supply and opposite side lower-exhaust scheme				Up-supply and up-return with high efficiency air cleaner (CDC from U.S.A.)	Up-supply and two-side return (CDC from U.S.A.)
	Single deflection grille	Double shutter	Square ceiling diffuser	Low-velocity perforated ceiling supply air outlet		
Air velocity near the face, m/s	0.05	0.05	0.09	0.05	0.03	0.13
PD,%	0	0	4.5	0	0	6.8

Fig. 4.36 Dispersion of airflow indoors with the double shutter

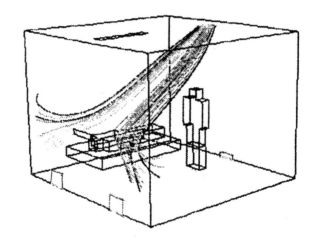

Fig. 4.37 Dispersion of droplet indoors from coughing of patient with the double shutter

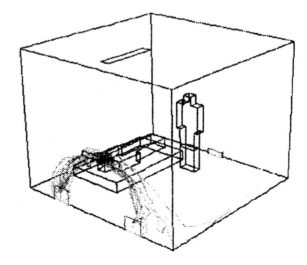

Table 4.7 Relationship between air velocity at air supply outlet and that near patient face

Air change rate, h^{-1}	Number of air supply outlets	Air velocity at air supply outlet, m/s	Air velocity at height 0.8 m, m/s
10	1	0.5	0.11
10	2	0.5	0.055
15	2	0.75	0.095
25	3	0.40	0.109
25	4	0.40	0.073

the performance of air distribution is the best. In this case, the air distribution indoors is shown in Fig. 4.36. The schematic diagram for the dispersion of pollutant indoors from the coughing process of patient is shown in Fig. 4.37 [10].

From all the figures shown in Sect. 4.3, when the air supply outlet is not above the face of patient, the relationship between the air velocity at the height 0.8 m and the air velocity at the air supply outlet can be obtained, which is shown in Table 4.7. With the conditions in the table, the air velocity near the face can be less than 0.1 m/s.

4.6.4 Experimental Validation on Performance of Opening Position

1. Experimental method

The most effective and intuitive way to validate the air distribution is to perform microbiological measurement. Figure 4.38 shows the layout of the isolation ward in experiment. Figure 4.39 shows the profile of the air supply and exhaust openings.

The effect of the air supply and exhaust openings is as follows.

(1) Air supply or return opening with self-purification (No. 1, 4 and 5 in the figure). It can guaranteed the room to perform self-purification rapidly, and the background concentration can be the air cleanliness level 1000. In this way, the measurement error can be reduced.

(2) The combination of air supply outlets A, B and C, together with air return (exhaust) openings a and b, forms the aforementioned four schemes in Sect. 4.6.3.

2. Experimental result

(1) Fig. 4.40 shows the sampling positions of the precipitating bacteria in the isolation ward. For sampling the bacterial concentration on bed, the sampling positions were on the bed, which is shown in Fig. 4.41. Other sampling positions were placed on the floor.

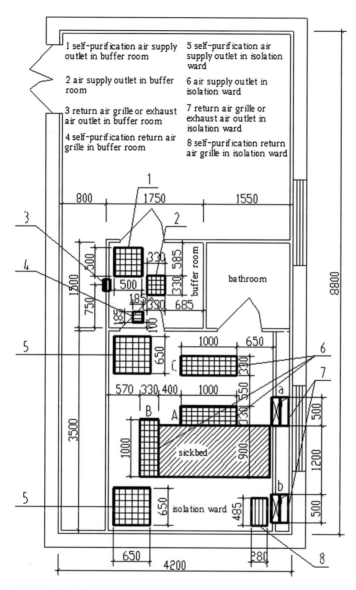

Fig. 4.38 Layout of the air supply and exhaust openings in simulated isolation ward

(2) The distribution of the mixed bacteria is presented in Fig. 4.42.

From Fig. 4.42, for scheme No. 1 (Air supply outlet B with air return openings a and b), the average concentration of the mixed bacteria is 1.1 CFU/petri dish. For scheme No. 2-a (Air supply outlet A with air return opening b), the average concentration of the mixed bacteria is 1.6 CFU/petri dish. For scheme

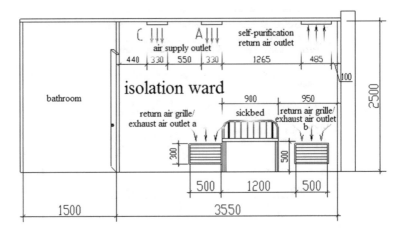

Fig. 4.39 Profile of the simulated isolation ward

No. 4 (Air supply outlet A and B with air return opening b), the average concentration of the mixed bacteria is 1.3 CFU/petri dish.

(3) The concentration of the experimental bacteria for different schemes in the isolation ward is shown in Fig. 4.43.

(4) The distribution for the concentration of the settlement bacteria in the region facing the operator is shown in Fig. 4.44.

3. Analysis

(1) At first, three kinds of the upper-supply and lower-exhaust schemes play the same role in the cleaning of indoor air. Based on the data of the mixed bacteria shown in Fig. 4.42, the difference is not obvious for the concentration range 1.1–1.6 CFU/Petri dish. Moreover, under the condition of the air change rate 12 h^{-1} (which could corresponds to the air cleanliness level 100,000), the bacterial concentration reached the value within air cleanliness level from 1000 to 10,000. This means the isolation ward has a certain capacity of air cleaning.

(2) Based on the principle of protecting the medical personnel, the experimental result for scheme No. 2-a (with one air return opening) is quite prominent.

The area of the sampling positions shown in the rectangular frame shown in Fig. 4.40 is in front of the patient, which includes the sampling points 22–25 and 27–30.

In this area, the average bacterial concentration for scheme No. 2-a (with one air return opening) is 8172 CFU/Petri dish. The average bacterial concentration for scheme No. 1 is 10007 CFU/Petri dish. The concentration for scheme No. 1 is larger than that for scheme No. 2-a (with one air return opening) by

(a)

(b)

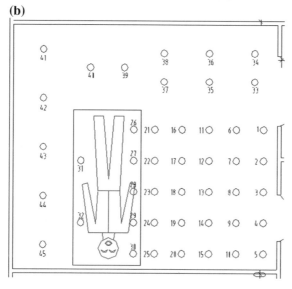

Fig. 4.40 Schematic of the sampling positions. **a** Photo-of-Petri-dish-on-floor. **b** Location-of-sampling-positions-in-isolation-ward

22.5%. For sampling points 22–30, the concentration for scheme No. 1 is larger than that for scheme No. 2-a (with one air return opening) by 25%. Therefore, it is meaningful to adopt scheme No. 2-a (with one air return opening) in this aspect. As mentioned before, this result has been proved by numerical simulation.

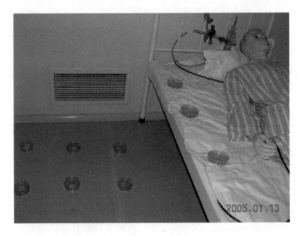

Fig. 4.41 Schematic of the sampling positions on the bed

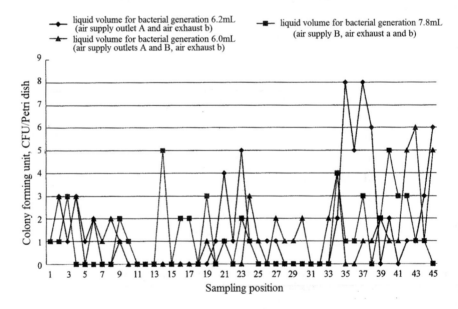

Fig. 4.42 Distribution of the mixed bacteria in the isolation ward

Although there is no experimental data for scheme No. 2-b with two air return openings and scheme No. 3, it is shown from numerical result shown in Table 4.4 that the concentrations in the breathing zone at the height 0.65–1.5 m for scheme No. 2-b with two air return openings and scheme No. 3 are larger than scheme No. 2-a with one air return opening and scheme No. 4.

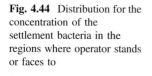

— ◆ — liquid volume for bacterial generation 6.2mL   — ■ — liquid volume for bacterial generation 7.8mL
         (air supply A, air exhaust B)   (air supply B, air exhaust A and B)
— ▲ — liquid volume for bacterial generation 6.0mL
         (air supply A and B, air exhaust B)

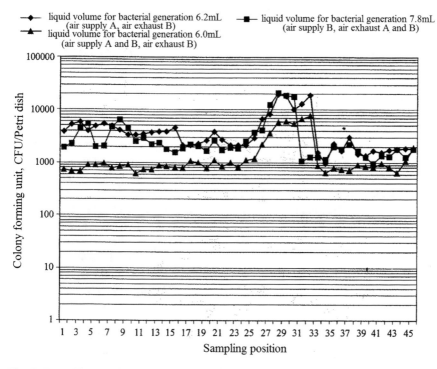

Fig. 4.43 Distribution of the actual experimental bacteria with different schemes in the ward ($n = 12$ h^{-1}, the concentration of the bacterial solution is 8×10^{-10} pc/mL)

Fig. 4.44 Distribution for the concentration of the settlement bacteria in the regions where operator stands or faces to

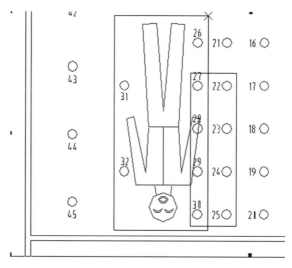

It has been mentioned before that it is not important how much is the average indoor concentration, but the concentration in the working area of the medical personnel should be lowest. Therefore, the performance of scheme No. 2-a with one air return opening is superior to that of scheme No. 1, scheme No. 2-b with two air return openings and scheme No. 3.

(3) Why the concentrations for sampling points 26–32 on the bed for various schemes are obviously larger than other sampling points? The concentration of scheme No. 2-a with one air return opening is larger than the indoor average concentration by two times, which is also larger than that for scheme No. 1 by three times and larger than that for scheme No. 4 by three and four times. This is because these petri dishes were placed on the bed, which is close to the pollutant source. This is understandable.

(4) The whole performance measured for scheme No. 4 is superior to that of scheme No. 2-a with one return air opening and scheme 1, while the performances of the latter two schemes are better than other schemes by both measurement and simulation. Based on the simulated result by Weipeng Deng shown in Table 4.4, the performance for the scheme with one-side top-supply is the best. Therefore, it is easy to combine these findings for understanding purpose. Because of the advantages of these two schemes, they are widely used.

(5) As mentioned in Sect. 3.3, the area of the air supply outlet can be enlarged until the air velocity at the air supply outlet is not less than 0.13 m/s.

(6) Two air supply outlets can be placed separately as the scheme No. 4, on condition that the air supply velocity is not less than 0.13 m/s.

References

1. F. Chan, V. Cheung, Y. Li, A. Wong, R. Yau, L. Yang, Air distribution design in a SARS ward with multiple beds. Build. Energy Environ. **23**(1), 21–33 (2004)
2. B.B. Батурин, Y. Liu, *Fundamentals of Industrial Ventilation* (China Industry Press, Beijing, 1965), p. 109
3. Y. Sun, *Industrial Ventilation*, 2nd edn. (China Architecture & Building Press, Beijing, 1985)
4. Z. Xu, Y. Zhang, Y. Zhang, Z. Mei, J. Shen, D. Guo, P. Jiang, H. Liu, Mechanism and performance of an air distribution pattern in clean spaces. J. HV&AC **30**(3), 1–7 (2000)
5. Y. Zhang, Z. Xu, Y. Zhang, Z. Mei, J. Shen, D. Guo, P. Jiang, Numerical simulation analysis on concentration field in clean room with different air supply areas. Build. Sci. **15**(6), 6–11 (1999)
6. H. Liu, Analysis on factors influencing air distribution of local clean zone with air cleanliness level 100 in clean space. China Academy of Building Research (2000)
7. Z. Xu, *Fundamentals of Air Cleaning Technology* (Springer, Berlin, 2014), p. 293
8. Z. Xu, *Fundamentals of Air Cleaning Technology* (Springer, Berlin, 2014), p. 311
9. Z. Xu, Y. Zhang, Y. Zhang, X. Yu, Discussion on the place of supply air outlet and return air inlet in biosafety laboratories. Contam. Control Air-Conditioning Technol. **4**, 15–20 (2005)

10. W. Deng, The synthetic control measures and strategies for preventing the transmission and infection of SARS in hospitals. Tongji University (2005)
11. Z. Xu, Y. Zhang, Q. Wang, F. Wen, H. Liu, L. Zhao, X. Feng, Y. Zhang, R. Wang, W. Niu, Y. Di, X. Yu, X. Yi, Y. Ou, W. Lu, Study on isolation effects of isolation wards (1). J. HV&AC **36**(3), 1–9 (2006)
12. X. Feng, Z. Xu, Y. Zhang, Q. Wang, H. Liu, F. Wen, X. Yu, L. Zhao, R. Wang, Y. Zhang, W. Niu, X. Yi, Y. Ou, W. Lu, Analyses of numerical simulation and the effect on air distribution of negative pressure isolation rooms. Build. Sci. **22**(1), 35–41+45 (2006)
13. J. Shen, Multi-application isolation ward and its air conditioning technique without condensed water. Build. Energy Environ. **24**(3), 22–26 (2005)
14. X. Tang, J. Shen, W. Deng, C. Li, Effects of supply-air outlet on unidirectional air distribution in contagious isolation wards. Build. Energy Environ. **24**(4), 11–18 (2004)

Chapter 5
Calculation of Air Change Rate

5.1 Outline

In order to dilute and effectively remove the bioaerosol in the negative pressure isolation ward, a certain amount of flow rate, i.e., the air change rate, is needed. According to the principle of air cleaning technology, the performance with larger air change rate is better. This has been proved and no further experimental verification is needed. For example, it is of no use to prove that the performance with the air change rate $12 \, h^{-1}$ is better than that with $10 \, h^{-1}$, or the performance with $10 \, h^{-1}$ is better than that with $8 \, h^{-1}$. But the relationship is not completely linear proportional. It does not mean that better performance will be obtained when the air change rate increases too much. There is an economic problem.

For the certain user such as hospital, too much air change rate cannot be beard. Then how much is it suitable? Just as that mentioned in Chap. 1, it is almost blank in the research and construction of isolation ward these years. Therefore, the related material is very rare. It is difficult to verify and summarize through practice, not to mention that the comprehensive practice is difficult to be found.

So usually the specification or suggestion from related literatures from U.S.A. are used as reference. For example, in ASHRAE manual in U.S.A., based on the requirement of thermal comfort, the air change rate in the isolation ward for the pulmonary tuberculosis patient is $6 \, h^{-1}$. This is much less than that specified in other literatures, which is shown in Table 5.1.

Since there is no fundamental answer based on theory, the determination for the value of the air change rate is still controversial.

Then how could we determine the reasonable air change rate? Is it possible to use the isolation coefficient based on the biosafety cabinet to obtain the allowable microbial standard in the isolation, so that the problem for determining the air change rate in isolation ward can be solved? It is obviously infeasible. The safety limit outside the biosafety cabinet is determined under the condition that operator is standing in front of a biosafety cabinet with the isolation effect. But the bioaerosol

© Springer Nature Singapore Pte Ltd. 2017
Z. Xu and B. Zhou, *Dynamic Isolation Technologies in Negative Pressure Isolation Wards*, DOI 10.1007/978-981-10-2923-3_5

Table 5.1 Related standards on the flow rates of the dilution air and the outdoor air

Standard	Specified flow rates of the dilution air and the outdoor air
"Guidelines for Preventing the Transmission of Mycobacterium tuberculosis in Healthcare Facilities" issued by CDC in U.S.A.	In new-built or renovated isolation ward for prevention of airborne transmission, the air change rate >12 h^{-1} and the outdoor air volume >2 h^{-1}
ASHRAE manual (2003) "Health Care Facilities" [1]	In isolation ward, the flow rate of dilution air >6 h^{-1} (based on requirement for odor and thermal comfort)
UK "Guidance on the prevention and control of transmission of multiple drug-resistant Tuberculosis" [1]	In new-built or renovated isolation ward for prevention of airborne transmission, the flow rate of dilution air $\geq$ 12 h^{-1} and the outdoor air volume $\geq$ 2 h^{-1}
CDC in U.S.A. "Guidelines for environmental infection control in health care facilities" [1]	In newly-built ward, the air supply volume $\geq$ 12 h^{-1}. In existing ward, the air supply volume $\geq$ 6 h^{-1}.
AIA in U.S.A. "Guidelines for design and construction of hospital and health care facilities" [1]	In isolation ward for prevention of airborne transmission, consulting rooms for emergency or radiotherapy, the flow rate of dilution air $\geq$ 12 h^{-1} and the fresh air volume $\geq$ 2 h^{-1}
DHHS in U.S.A. "Guidelines for construction and equipment of hospital and medical facilities" [1]	In isolation ward for prevention of airborne transmission, the air supply volume $\geq$ 12 h^{-1} and the fresh air volume $\geq$ 2 h^{-1}. In the bathroom, laundry, waste disposal room, disinfection room, anteroom of isolation ward, the exhaust air volume $\geq$ 10 h^{-1}
ASHRAE 170-2013 "Ventilation of Health Care Facilities"	In new-built isolation ward for prevention of airborne transmission, the air change rate $\geq$ 12 h^{-1} and the outdoor air volume $\geq$ 2 h^{-1}. In the antechamber, the air change rate for exhaust air $\geq$ 10 h^{-1}
Australia "Guidelines for the classification and design of isolation rooms in health care facilities" [1]	In negative pressure isolation ward, the flow rate of dilution air should be the larger value between 12 h^{-1} and 522 m^3/h
USA "Guidelines on the design and operation of HVAC systems in disease isolation areas" [1]	In newly-built isolation ward, disposal room and mortuary, the flow rate of dilution air $\geq$ 12 h^{-1}. In the bathroom, the exhaust air volume $\geq$ 10 h^{-1}. In the consulting room for infectious patient, the flow rate of dilution air $\geq$ 15 h^{-1} and the fresh air volume $\geq$ 2 h^{-1}
"Guideline for Design and Operation of Hospital HVAC Systems" (HEAS-02-2004) established by Healthcare Engineering Association of Japan	In isolation ward, the total volume should of exhaust air be 12 h^{-1}

in the isolation ward disperses in the whole room. Therefore, it is difficult to provide the requirement in the isolation ward. Moreover, the bacterial generation rate cannot be determined based on one kind of disease.

In this chapter, one thought to calculate the air change rate in the negative pressure isolation ward will be proposed in a comprehensive way.

5.2 Two System Modes of Isolation Ward

5.2.1 Circulation Air System

In terms of air cleaning handling, the circulation air system is shown in Fig. 5.1. In the figure, N_t is the indoor particle concentration at time t, pc/L; N is the stable particle concentration indoors, pc/L; N_0 is the initial particle concentration indoors, which is the particle concentration at time 0, pc/L; V is the volume of the clean-room, m^3; N is the air change rate, h^{-1}; G is the particle generation rate per unit volume indoors, pc/(m^3 min); M is the concentration of atmospheric dust, pc/L; S is the ratio of the return air volume to the supply air volume.

When the purpose is to remove bioaerosols from indoor air, the implications of η in the figure are: η_1 is the efficiency of the coarse filter (or the combination of air filter for the outdoor air) for bioaerosol (the particle counting efficiency, which is expressed with decimal); η_2 is the efficiency of the middle filter for bioaerosol; η_3 is the efficiency of the final filter for bioaerosol.

Under the steady-state condition, the calculation equation for the indoor bacterial concentration is:

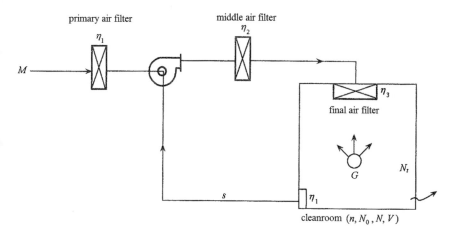

Fig. 5.1 Schematic diagram of circulation air system

$$N = \frac{60G \times 10^{-3} + Mn(1-s)(1-\eta_n)}{n[1-s(1-\eta_r)]} \qquad (5.1)$$

When there is no specific bioaerosol from the supply air into the isolation ward, $M = 0$. When HEPA filter is installed on the return air grille or the air supply outlet, or both, the efficiency for bioaerosol reaches more than 99.999%. Equation (5.1) can be simplified as:

$$N = \frac{60G \times 10^{-3}}{n} \qquad (5.2)$$

$$n = \frac{60G \times 10^{-3}}{N} \qquad (5.3)$$

5.2.2 Full Fresh Air System

For the full fresh air system, there is only supplied air and exhausted air. The flow rate of exhausted air is larger than that of the supplied air. The schematic diagram is shown in Fig. 5.2. Because $s = 0$, we obtain

$$N = \frac{60G \times 10^{-3} + Mn(1-s)(1-\eta_n)}{n} \qquad (5.4)$$

With the same method, Eqs. (5.2) and (5.3) can be also simplified. This means that no matter what kind of the system is, as long as there is no such kind of dust source or bacterial source in the atmosphere and HEPA filters are installed for the exhausted air or the return air, the equation to calculate the air change rate is the same.

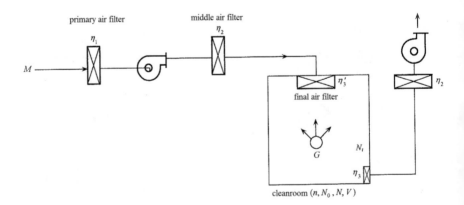

Fig. 5.2 Schematic diagram of the full fresh air system

From Eq. (5.3), two parameters must be known for calculating the air change rate, which are the bacterial generation rate indoors and the standard for the bacterial concentration.

5.3 Determination of Bacterial Generation Rate Indoors

5.3.1 Bacterial Generation Rate from Ordinary Patients

In Eqs. (5.1)–(5.3), the value of G should be determined at first. G is the bacterial generation rate per unit volume. The bacterial generation rate per unit time should be known, so that under the assumption of the uniform distribution, the bacterial generation rate per unit volume indoors G could be obtained. Then after considering the non-uniform distribution, the air change rate n can be obtained with the non-uniform distribution coefficient Ψ.

As mentioned before, it is not comprehensive to use the data on the bacterial generation rate for individual disease which occurred occasionally. Based on the analysis on foreign literates [2], we could obtain the following information.

1. When people wear the asepsis clothes, the bacterial generation rate in the static state is 100–300 pc/(min people). When the common activity is carried out on body, the bacterial generation rate is 150–1000 pc/(min people). When people walk fast, the bacterial generation rate is 600–2500 pc/(min people).
2. When people wear the ordinary cloth, the bacterial generation rate is 300–62000 pc/(min people).
3. During the coughing process for once time, the bacterial generation rate is 70–700 pc/(min people). During the sneezing process for once time, the bacterial generation rate is 4000–60000 pc/(min people), which is slightly different from that in Fig. 1.3.
4. The ratio of the bacterial generation rate between the condition when people wear mask and that without mask is 1:7–1:14.
5. The ratio of the bacterial generation rate to the particle generation rate is 1:500–1:1000.

5.3.2 Analysis of Bacterial Generation Rate from Respiratory System

In previous section, the bacterial generation rates during coughing and sneezing processes are given. But the information on bioaerosol size is not given. The frequency of sneezing per hour is unknown.

The data in detail was given in Fig. 1.3. When people sneeze for once time, the generation rate of aerosol with diameter 10 μm is 3×10^5 pc, which is larger than the value given in the previous section. The generation rates of aerosol with diameter 1 and 0.3 μm are 2×10^4 pc and 10 pc, respectively. When people cough for once time, the generation rates of aerosol with diameter 10, 1 and 0.3 μm are 1. 5×10^3 pc, 30 pc and 1 pc, respectively.

However, the droplets generated during coughing and sneezing processes will evaporate quickly. The evaporation rate is related to the temperature, the vapor pressure of the droplet and the atmospheric pressure. The evaporation time of the droplet can be obtained with the following equation [3]:

$$
t = -\left\{ \frac{D_p R T \rho_L}{2 \alpha v (\delta P_1 - P) M_1} + \frac{R T \rho_L}{P_1 (\delta P_1 - P) M_1} \left[\frac{D_P^2}{8} \cdot \frac{D_P}{2} \lambda + \lambda^2 \log\left(1 + \frac{D_P}{2\lambda}\right) \right] \right\}
$$

(5.5)

According to the calculation performed by Chen [4], the following values can be chosen for various parameters of water droplet.

t is the time, s; D_p is the diameter of the droplet, which is 9.53×10^{-4} cm; R is the gas constant, which is 8.317×10^7 erg/(mol K); T is the absolute temperature, which is 293 K (i.e., 20 °C); ρ_L is the density of the liquid, which is 1 g/cm^3 for H_2O; α is the evaporation coefficient, which is 0.04 for H_2O; $v = \left(\frac{RT}{2\pi M_1}\right)^{1/2}$ is the kinematic viscosity, cm^2/s; M_1 is the molecular mass of the liquid vapor, which is 18 g/mol for H_2O; δ is the degree of saturation, which is 0.3 (this means the relative humidity is 30%); P_1 is the vapor pressure of the droplet (here it is the differential pressure of the water vapor), which is $5.6 \times 1.33 \times 10^3$ dyn/cm^2; P is the vapor pressure of the droplet, which is $5.6 \times 1.33 \times 10^3$ dyn/cm^2; D_1 is the dispersion coefficient of the liquid vapor through the gas, which is 0.252 cm^2/s; λ is the average mean free path of the molecular, which is 6.53×10^{-6} cm.

For the water droplet with diameter 9.5 μm, it can be completely evaporated in 0.104 s. When the relative humidity is 60%, it only takes 0.193 s to evaporate completely.

According to the calculated results by foreign scholars, under the condition of relative humidity 70%, the water droplet with diameter 5 μm will evaporate in 0.029 s [5], which is much faster than the above result.

On contrary to the water, the salive is the secretion from the parotid gland, the submandibular gland, the sublingual gland and the minor salivary glands. According to the analysis by Kou, water component occupies more than 99% of the salive. The solid component accounts for about 0.7%, where the organic compound and the inorganic compound occupy 05 and 0.2%, respectively [6]. Moreover, 0.85% of the body fluid is NaCl on average, which is also the standard value for the normal saline. Therefore, the maximum ratio of the NaCl in the salive is the level of the body liquid.

During the evaporation process, the droplet of the salive will be concentrated gradually until the phase change process occurs. In this case, the values of various parameters are not the same as that of the water droplet. Therefore, the time for evaporation will be prolonged.

As we know, the test aerosol used for HEPA filter and medical mask is the sprayed NaCl, which is generated from the NaCl solution with concentrate 2%. According to British test standard [7], Chinese literature [8] and the test standard for HEPA filter in China, the time of evaporation for sprayed NaCl droplet is 2 s.

Another problem is that the salive containing the salt component cannot evaporate to vanish completely as the water. The solute will stay airborne as the small salt particle. This means even when the salive droplet will evaporate thoroughly, it will become the extremely small particle. When the salive contains the bacteria, the particle with similar size of the bacteria will be left.

Because of the dehydration, bacteria will move and evaporate with the water molecular, which will change the property of the protein in the bacteria and may cause the death of the bacteria. Table 5.2 shows the recovery rate of the bacteria during the spraying process of droplet [9]. It is shown that the recovery rate of water during the spraying process is the maximum.

The solute particle left after evaporation of droplet may be larger than the size of the bacteria it contains, or may be smaller than the size of the bacteria. Suppose

Table 5.2 Recovery rate of different bioaerosols

Type of aerosol	Component of spray fluid	Component of sampling fluid	Average recovery rate, %	Times of experiments
Phenolsulfonphthalein	Water	Water	86.9	2
Bacillus subtilis var. niger	Water	Water	73.4	3
Bacterium prodigiosum	2 gelatin aqueous solution	Gelatin dilution solution	25.1	6
Brucella	Tryptose sline containing 5% maltodextrin	Tryptose saline	25.9	5
Burkholderia pseudomallei	2% glycerol broth	Broth	16.6	15
Malleomyces Mallei	2% glycerol broth	Broth	10.1	18
Francisella tularensis	2% glycerol	1% gelatin saline	2.3	34
Meningitis pneumonia virus	Broth	Broth	9.5	7
Psittacosis virus (6BC)	Broth	Broth	18.1	19
Psittacosis virus (3Drg)	Broth	Broth	23.0	3

both the droplet and the solute particles are spherical, the relationship between their sizes can be expressed as follows [3],

$$d_P = d_D\left(\frac{c\rho_D}{\rho_P}\right)^{1/3} \tag{5.6}$$

Where d_P is the diameter of the solute particle, μm; d_D is the diameter of the droplet, μm; c is the percentage for the mass of the solute in the liquid droplet, which is set 0.85%; ρ_P is the density of the solute particle, which is 2.164 g/cm^3 for NaCl; ρ_D is the density of the liquid droplet, which could be approximated as the density of water 1 g/cm^3 since the value of c is extremely small.

Therefore, the size of the pure solute particle can be calculated as follows.

$$d_P = d_D\left(\frac{0.0085 \times 1}{2.164}\right)^{1/3} = 0.16d_D$$

For the droplet with diameter 10 μm, its smallest size is not smaller than 1.6 μm after 2 s of evaporation. For the droplet with diameter 5 μm, its smallest size is not smaller than 0.8 μm after 2 s of evaporation. For the droplet with diameter 1 μm, its smallest size is not smaller than 0.16 μm after 2 s of evaporation. For the droplet with diameter 0.5 μm, its smallest size is not smaller than 0.08 μm after 2 s of evaporation.

By preliminary investigation, it is found that coughing will easily occur in the evening and in the morning when people get cold. In the morning, the average frequency of coughing is one time per 2 s. After getting up in the morning, it is five times per hour. When the maximum bacterial generation rate per hour is calculated based on this data (the frequency per hour is not always such large), the result will have a quite large deviation. After 2 s of evaporation, we obtain:

For the size 1.6 μm,

$$\frac{5 \times 2.5 \times 10^5 + 30 \times 1.5 \times 10^3}{60 \times 25} = 863 \, \text{pc/(m}^3 \, \text{min)}$$

For the size 0.8 μm,

$$\frac{5 \times 6 \times 10^4 + 30 \times 200}{60 \times 25} = 204 \, \text{pc/(m}^3 \, \text{min)}$$

For the size 0.16 μm,

$$\frac{5 \times 2 \times 10^4 + 30 \times 30}{60 \times 25} = 66.7 \, \text{pc/(m}^3 \, \text{min)}$$

For the size 0.08 μm,

$$\frac{5 \times 30 + 30 \times 1}{60 \times 25} = 0.12 \, \text{pc/(m}^3 \, \text{min)}$$

Now Eq. (5.6) is applied to analyze the final size of the sprayed droplet.

It is known that the percentage for the mass of NaCl in the sprayed droplet is 0.85%, and the main diameter of the droplets is 5 μm, the diameter of the solute particles after complete evaporation can be calculated as follows:

$$d_P = 5 \times \left(\frac{0.0085 \times 1}{2.164}\right)^{1/3} = 5 \times 0.16 = 0.8 \, \mu m$$

This diameter value is close to that of the Bacillus subtilis as shown in Fig. 3.21.

5.4 Determination of Bacterial Concentration Standard Indoors

5.4.1 Outline

For obtaining the air change rate, the standard for the allowable bacterial concentration indoors must be known. But for different diseases, the pathogenic concentrations are obviously different. And for patients and medical personnel, the concentrations are also different.

Therefore, it is difficult to set the standard for the pathogenic dosage of indoor air with specific disease or microbes. Take the animal experiment as an example, when experiment with streptococcus and influenza virus as test aerosol on rabbit, results can be obtained as shown in Table 5.3 [10].

In our opinion, the influence of the pathogenic microbes generated during speaking and coughing process from patient on the patient himself should not be considered. This is why several patients with pulmonary tuberculosis can live in the same room. Therefore, it is of first priority to consider the influence on the medical personnel.

Fundamental protective measures are provided for the medical personnel entering into the isolation ward. The main measures are wearing N95 masks.

Table 5.3 Median lethal dose of pathogenic microbes

Pathogenic microbes	Streptococcus			Influenza virus, PFU		
Test method	Aerosol	Nasal drop	Respiratory infection	Aerosol	Nasal drop	Respiratory infection
Median lethal dose, LD_{50}	$10^{4.95}$– $10^{7.41}$	$10^{3.76}$– $10^{4.13}$	$10^{3.87}$– $10^{4.37}$	$10^{1.46}$– $10^{1.86}$	$10^{1.06}$– $10^{1.80}$	$10^{1.21}$

Table 5.4 Efficiency of homemade polypropylene fibrous media for DOP, %

ΔP	20.3 Pa	40.6 Pa	59.9 Pa	116.9 Pa	482.6 Pa
v	1.25 m/s	2.5 m/s	3.5 m/s	7.0 m/s	13.4 m/s
0.5 μm	–	–	–	–	99.83
0.4 μm	99.9994	99.995	99.984	99.83	99.59
0.3 μm	99.9991	99.990	99.970	99.71	99.12
0.2 μm	99.9950	99.970	99.920	99.41	98.37
0.15 μm	99.9960	99.870	99.780	98.48	97.78
0.10 μm	99.8800	99.660	99.370	98.62	96.26
0.08 μm	99.8500	99.560	99.270	96.52	95.39
0.05 μm	99.8400	99.400	99.170	95.54	93.80
0.03 μm	99.9700	99.860	99.710	98.92	–

In the early of May in 2003 during the epidemic of SARS, according to the TV report at Hong Kong, there were no people infected in a hospital where all the medical personnel wore N95 masks. However, for these hospitals where the ordinary masks were worn, many medical personnel were infected.

The filtration efficiency of N95 mask for particles with diameter 0.075 μm reached more than 95%, which will be introduced in detail later. From Table 5.4 [11], the efficiency of the mask made of polypropylene fibrous media for particles with diameter 0.075 μm (which is close to 0.08 μm in the table) under the filtration velocity 13.4 cm/s is the lowest. It is safe when this value is set as the standard.

NIOSH standard 42CRF84 has pointed out that the N95 mask meets the standard requirement for mask to prevent the solid tuberculosis bacteria. N95 means the efficiency with the sodium flame method for particles with diameter 0.075 μm is 95%, which meets the standard requirement for N-type. In China, it was misunderstood in some literatures that N95 means the efficiency for DOP with diameter 0.3 μm is 95%. Moreover, there are also requirements for R-type mask with oil aerosol and P-type mask with both solid and liquid aerosol (they are based on the DOP efficiency).

After the epidemic of SARS, the standard GB2626 "*Respiratory protective equipment. Non-powered air-purifying particle respirator*" was revised. There are two types including KN-type and KP-type, which correspond with N-type and P-type in U.S.A. Detailed information is shown in Table 5.5.

Table 5.5 Classification of masks in China

Classification level of air filtration unit	Test with NaCl	Test with oil aerosol
KN90	≥ 90.0%	Not applicable
KN95	≥ 95.0%	
KN100	≥ 99.97%	
KP90	Not applicable	≥ 90.0%
KP95		≥ 95.0%
KP100		≥ 99.97%

The test conditions in Chinese standard are the same as that in U.S.A. The concentration of NaCl is 2% with the median diameter 0.075 μm ± 0.02 μm. The flow rate is 85 ± 4 L/min. There is no requirement for the size of the mask, but it is usually 100 cm². The corresponding filtration velocity is 14.17 cm/s.

5.4.2 Standard

Therefore, there are two aspects to consider the standard value of the bacteria indoors. One is from the medical personnel. The other is from the environment.

1. From the medical personnel

   (1) Standard for droplet nuclei with diameter 0.075 μm
       According to the previous section, the filtration velocity can be calculated based on the flow rate and the size of the mask, which is 14.17 cm/s. The corresponding maximum penetration particulate size is equivalent with 0.075 μm. This means that the particle size 0.075 μm corresponds to the minimum efficiency under the filtration velocity 14.17 cm/s. Therefore, test aerosol with diameter 0.075 μm and the filtration velocity 14.17 cm/s are the most unfavorable conditions. When the efficiency under the most unfavorable conditions is met, the requirement under other conditions can be easily satisfied.
       As mentioned before, there were no one infected when they wore N95 masks. From the aspect of most stringent requirement, there was no bioaerosol inhaled. If even one particle with diameter 0.075 μm was inhaled, there is no possibility for other droplet nuclei to be inhaled (since the efficiency with other particles is higher than that with 0.075 μm).
       When the medical personnel enter into the ward, it usually takes less than 0.5 h. When special treatment is needed, it may take more than 1 h. In this case, it can be assumed 1.5 h.
       According to Chinese standard on medical masks, the requirement for the efficiency 95% is under the flow rate 85L/min. We know that the common respiratory flow rate is between 8L/min and 80L/min. It is a much safer consideration for the standard to set such flow rate.
       When the flow rate of respiratory air is 85L/min, in the inhaled air within 1.5 h, the concentration of particles with diameter 0.075 μm will not be larger than $1/(90 \times 85) = 1.3 \times 10^{-4}$ pc/L. Therefore, when the efficiency of mask reaches 95%, the concentration of particles with diameter 0.075 μm should not exceed 2.6×10^{-4} pc/L.
   (2) Standard for droplet nuclei with diameter 0.075 μm
       From Fig. 1.3, it is known that the number of airborne droplet with diameter 10 μm is the most. From the Sect. 5.3.2, the diameter of the airborne nuclei is 1.6 μm.

The concentration of particles with diameter 1.6 μm in the inhaled air should not exceed 1.3×10^{-4} pc/L.

When the N95 mask is worn, the efficiency for 1.6 μm is larger than that for 0.075 μm by three orders of magnitude. Suppose the efficiency is 99.998% (please refer to Chap. 3), the concentration of particles with diameter 1.6 μm in the inhaled air should not exceed 6.5pc/L.

(3) Standard for ordinary microbes indoors

When particles are captured by masks, the microbial particles are filtered. For the isolation ward where air is supplied through HEPA filter, the particle concentration is based on the diameter ≥ 0.5 μm. From Table 5.4, the efficiency of N95 mask for particles with diameter 0.3 μm can be 99.12%. So the calculated efficiency for diameter ≥ 0.5 μm is 99.97% [12].

The concentration of particles with diameter ≥ 0.5 μm in the inhaled air should not exceed 1.3×10^{-4} pc/L. So the concentration of particles with diameter ≥ 0.5 μm indoors should not exceed 0.433pc/L.

2. From the environment

From the environment, there should be no bioaerosol released from the isolation ward. For Three-Room-One-Buffer scheme, i.e., the ward-buffer-exterior or corridor, the final bacterial concentration, i.e., the inhaled bacteria during occupancy period which enters into the exterior room should be less than 1pc. The period for passing through the exterior room is short. There is ventilation in the exterior room. The decayed concentration in 1 min will be more than a half. So the concentration within the first 1 min is considered.

When the flow rate of respiratory air is 85 L/min, in the inhaled air within 1.5 h, the concentration of particles with diameter 0.075 μm will not be larger than 0.012 pc/L, so that the total number of inhaled bacteria can be less than 1. When the volume of the exterior room is 25 m^3 (the less the volume is, the more stringent the bacterial concentration in the ward should be), the allowable bacterial number in the exterior room should not exceed $0.012 \times 25 \times 10^3 = 300$ pc. In this case, the leakage flow rate into the exterior room during the opening of the door in the buffer room for once time is 1.62 m^3 (please refer to the chapter about the buffer room), where the number of the microbes should not be larger than 300 pc. Given $\beta_{3\cdot1} = 40$, the bacterial concentration in the isolation ward should be $300/1620 \times 40 = 7.4$ pc/L. (When the volume of the exterior room is 250 m^3, the allowable concentration can be 74 pc/L).

5.5 Calculation of Air Change Rate

5.5.1 Calculation Based on the Minimum Airborne Droplet Nuclei with Diameter 0.075 μm

With the assumption of uniform distribution, it can be calculated with Eq. (5.3), i.e.,

$$n = \frac{60G \times 10^{-3}}{N}$$

Where G is the generation rate of particles with diameter 0.08 μm (it is equivalent to 0.075 μm) which are obtained after evaporation from the droplet with diameter 0.5 μm. From the previous section, it is 0.12 pc/(m³ min). N is the allowable concentration of particles with diameter 0.075 μm indoors. From the previous section, it is 2.6×10^{-3} pc/L

Therefore, under the condition of uniform distribution, we obtain the air change rate:

$$n = \frac{60 \times 0.12 \times 10^{-3}}{2.6 \times 10^{-3}} = 2.8 \, \text{h}^{-1}$$

According to the non-uniform distribution theory, when the air change rate $n = 3 \, \text{h}^{-1}$, the non-uniform distribution coefficient ψ can be 4 [13]. So with the non-uniform distribution, $n_v = 4n = 4 \times 2.8 = 11.2 \, \text{h}^{-1}$. This means the air change rate should be 11.2 h⁻¹ at least.

5.5.2 Calculation Based on the Maximum Airborne Droplet Nuclei with Diameter from 10 μm to 1.6 μm After Evaporation

Suppose the generation rate of particles G with diameter 0.08 μm is 863 pc/(m³ min), and $N = 6.5$ pc/L, the air change rate with uniform distribution assumption is $n = \frac{60 \times 863 \times 10^{-3}}{6.5} = 8 \, \text{h}^{-1}$.

According to the non-uniform distribution theory, when the air change rate $n = 8 \, \text{h}^{-1}$, the non-uniform distribution coefficient ψ can be 1.5 [13]. So with the non-uniform distribution, $n_v = \psi n = 1.5 \times 8 = 12 \, \text{h}^{-1}$. This means the air change rate should be 12 h⁻¹ at least.

5.5.3 Calculation Based on the Ordinary Microbial Particles Indoors

According to Sect. 5.3.1, when aseptic cloth (the cloth on patient should be washed and disinfected) is worn and ordinary activity is performed, the maximum the generation rate of particles G is 1000 pc/(min·people). We obtain

$$G = 1000/25 = 40\,pc/(m^3 \cdot min).$$

Based on the aforementioned data, $N = 0.433$ pc/L. According to the uniform distribution, the air change rate is:

$$n = \frac{60 \times 40 \times 10^{-3}}{0.433} = 5.5\,h^{-1}$$

With the same method based on the non-uniform distribution theory, the non-uniform distribution coefficient ψ can be 2.25 according to the literature [13]. So with the non-uniform distribution, $n_v = \psi n = 2.25 \times 5.5 = 12.4\,h^{-1}$. This means the air change rate should be 12.4 h^{-1} at least.

5.5.4 Calculation Based on Environmental Standard

The largest value of the above three generation rates of particles G is 863 pc/ $(m^3\ min)$. $N = 7.4$ pc/L. According to the uniform distribution, the air change rate is:

$$n = \frac{60 \times 863 \times 10^{-3}}{7.4} = 7\,h^{-1}$$

With the same method based on the non-uniform distribution theory, the non-uniform distribution coefficient ψ can be 1.75 according to the literature [13]. So with the non-uniform distribution, $n_v = \psi n = 1.75 \times 7 = 12.2\,h^{-1}$. This means the air change rate should be 12.2 h^{-1} at least.

Based on the calculation results of the air change rate based on the above four standards, we obtain:

Based on the least droplet nuclei with diameter 0.075 μm, the calculated air change rate is 11.2 h^{-1};

Based on the most liquid droplets with diameter 10 μm, the calculated air change rate is 12 h^{-1};

Based on the common indoor microbes, the calculated air change rate is 12.4 h^{-1};

Based on the environmental protection outside of the isolation ward, the calculated air change rate is 12.2 h^{-1}.

The air change rate under these four conditions is between 11 and 12 h^{-1}. This provides many rational understanding for the recommended valued in foreign standards shown in Table 5.1. This also provides the basis for the foreign standards.

Since the conditions during calculation are relatively stringent, and the requirement for the indoor average concentration of the isolation ward is not so important as that for protecting the medical personnel, it is reasonable to choose 8–12 h^{-1}. For the isolation ward containing patients with seriously diseases, the relative large value can be adopted. On the contrary, the relative small value can be chosen. By the aforementioned analysis, the requirement can be satisfied when the air change rate is 10 h^{-1}. However, it is small if the lower limit is set 6 h^{-1}.

In the air change rate, the outdoor air volume can be 3–4 h^{-1}, which is larger than that under the normal condition. *"Guideline for Design and Operation of Hospital HVAC Systems"* (HEAS-02-2004) established by Healthcare Engineering Association of Japan has pointed out that the outdoor air volume needed for the isolation ward should be larger than the average outdoor air volume, because the outdoor air needed for the sick people is much larger than that for the healthy people.

References

1. J. Shen, Multi-application isolation ward and its air conditioning technique without condensed water. Build. Energy Environ. **24**(3), 22–26 (2005)
2. Z. Xu, *Design of Cleanroom* (Seismological Press, Beijing, 1994), p. 23
3. R. Dennis, *Handbook on Aerosols* (University Press of the Pacific, 1976), p. 54
4. C. Chen, W. Zhao, R. Guo, *Experimental Study on the Testing Apparatus with the Sodium Flame Method on HEPA Filter* (Beijing, 1983)
5. X. Yu, *Modern Air Microbiology* (People's Military Medical Press, Beijing, 2002), p. 57
6. L. Kou, *Basic Clinical Laboratory Science* (People's Medical Publishing House, Beijing, 2002)
7. BS3928, *Method for Sodium Flame Test for Air Filters* (U.K., 1969)
8. Q. Ji, *Test Apparatus of HEPA Filter with the Sodium Flame Method* (Atomic Energy Press, Beijing, 1981), p. 36
9. X. Yu, *Modern Air Microbiology* (People's Military Medical Press, Beijing, 2002), p. 58
10. X. Yu, *Modern Air Microbiology* (People's Military Medical Press, Beijing, 2002), p. 15
11. Z. Xu, *Fundamentals of Air Cleaning Technology and Its Application in Cleanrooms* (Springer, Berlin, 2014), p. 169
12. Z. Xu, *Fundamentals of Air Cleaning Technology and Its Application in Cleanrooms* (Springer, Berlin, 2014), pp. 205–207
13. Z. Xu, *Fundamentals of Air Cleaning Technology and Its Application in Cleanrooms* (Springer, Berlin, 2014), pp. 205–207

Chapter 6
Air Exhaust and Air Return Safely Are Necessary Conditions

6.1 Importance of Non-leakage Apparatus for Air Exhaust and Air Return

As mentioned before, because the efficiency of HEPA filter is extremely large, it makes the circulation of air possible. But when there is leakage in HEPA filter and its installation frame, the advantage of HEPA filter is losing, which is not allowed for indoor and outdoor environment.

The filter media of the HEPA filter is the paper. Although it may be complete before delivery from the factory, it may be destroyed before installation. Therefore, for HEPA filter which is used for filtering aerosol with high risk of hazardous level, the scanning leakage test must be performed, which is required by related standards.

The schematic diagram of the scanning leakage test is shown in Fig. 6.1. The scanning test with the reciprocal overlap routine near the possible leakage positions over the air supply surface of the filter and its installation frame is performed. The distance between the sampling probe and the downstream air supply surface of the filter should be less than 2.5 cm. The scanning pathway covers all the surfaces, the frame and the joint between the filter and the frame. The velocity of the scanning probe should be between 2 and 3 cm/s. A certain degree of overlap for the scanning pathway is allowed.

For the purpose of convenience, the scanning test with particle counting method for atmospheric dust can be adopted. The standard of the isolation ward can refer to that for Class Three biosafety laboratory [1]. When the indoor concentration of particles with diameter ≥ 0.5 μm is not smaller than 4000 pc/L (this can be obtained when the air supply is stop under the condition of door opening), the leakage of air filter can be considered if the downstream sampling concentration exceeds 3 pc/L. By applying the "Non-zero test principle", this kind of leakage test method with the particle counting means is much stricter and more accurate than that with the relative concentration [1].

© Springer Nature Singapore Pte Ltd. 2017
Z. Xu and B. Zhou, *Dynamic Isolation Technologies in Negative Pressure Isolation Wards*, DOI 10.1007/978-981-10-2923-3_6

Fig. 6.1 The schematic
diagram of the scanning
leakage test. *1* Frame; *2*
Scanning pathway; *3* Filter
media; *4* Separation plate

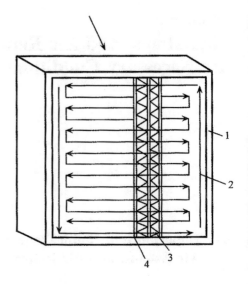

    Why the scanning leakage test is required for air filter, instead of the test for the
whole efficiency itself?
    As shown in Fig. 6.2, when one or several measurement points are placed close
to the upstream and downstream surfaces of the air filter, it is impossible to find the
leakage point on the edge. This is because the air flow at the downstream of the air
filter is unidirectional and in parallel. But both in theory and in practice, leakage
may exist on the installation frame or even on two sealing edges of the air filter.

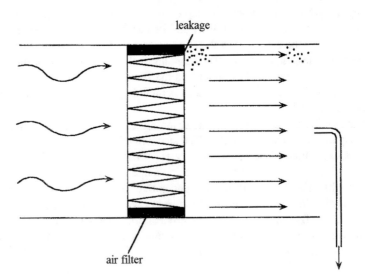

Fig. 6.2 The leakage test cannot be replaced by the efficiency test

Zero-leakage cannot be guaranteed by any mechanical sealing means. It can only reduce the extent of the leakage to the minimum under certain amount of the pressure. Meanwhile, it cannot guarantee that the leakage hole will never be expanded. Since the air filter is installed vertically at the air exhaust opening, the oil groove sealing means cannot be applied. Therefore, in order to be sure that the HEPA filter installed at the air return opening works well, the scanning leakage test must be performed after installation. After installation, the scanning can be only performed near the frame, because in-site leakage test can be performed on HEPA filter itself before installation.

   Because the scanning process should be performed at the air outlet surface and the air outlet of the HEPA filter for supplied air is within the spacious space (indoors or inside the equipment), the operation is feasible. But when air filter is installed at the air exhaust opening, which is connected with the air exhaust pipeline afterwards, the air outlet of the HEPA filter for exhaust air is in the confined space. Although by using the scanning test before installation can prove that the air filter itself has no leakage, the scanning leakage operation cannot be performed after installation if there is leakage on the frame, which is shown in Fig. 6.3.

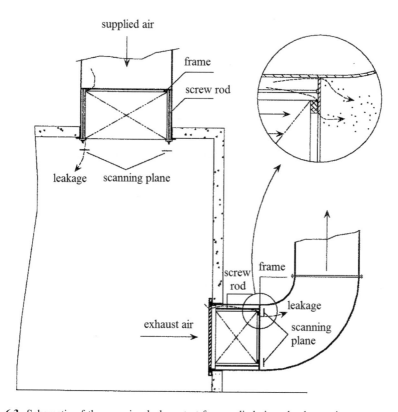

Fig. 6.3 Schematic of the scanning leakage test for supplied air and exhaust air

Fig. 6.4 Installation of air filter for exhaust air on the double wall

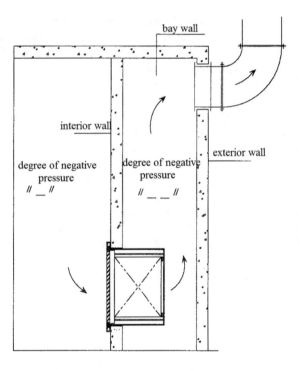

In some applications, the double wall is designed, which is shown in Fig. 6.4. In this case, people may enter into the double wall for the scanning test at the air outlet of the air filter. However, the shortcomings include:

(1) It increases the floor area and the construction cost. Separate double walls must be constructed for various rooms.
(2) The magnitude of the negative pressure inside the double wall is higher than that of indoors by tens of times, or even hundreds of Pascal. For the interior wall, especially those assembled from the color steel plate, cannot bear such high strength. And it is difficult to guarantee the air-tightness performance. Although the problem of air-tightness can be solved by the leakage test during the inspection stage, pollution will be caused inside the double wall once the problem of the air-tightness occurs again.

Therefore, as long as the negative pressure high efficiency air exhaust device is leakage-free, it becomes the important aspect to guarantee the performance of the isolation ward when circulation air is used.

6.2 Principle of Negative Pressure and Effective Air Exhaust Device Sealed with Dynamic Air Current

6.2.1 Structure of the Device

Because for any traditional mechanical sealing means or glue sealing, there is no exceptional absolute air-tightness in theory and in practice. By applying the principle of the air flowing from the side with high pressure towards the other side with low pressure, a negative pressure high efficiency air exhaust equipment with dynamic air current was invented, which is shown in Fig. 6.5. It is a novel idea in the field of sealing.

By special structural design, the sealing cavity containing airflow with positive pressure is formed between the shell of the air exhaust device and the shell of the air filter. Air with positive pressure can only flow out of the cavity. Part of the air flows towards into the room, and the other part will flow towards the air exhaust pipeline. The air with negative pressure, which contains the bacteria from the indoor air, will not flow into the cavity and then flow towards the air exhaust pipeline to the ambient. This equipment is the patented product of China.

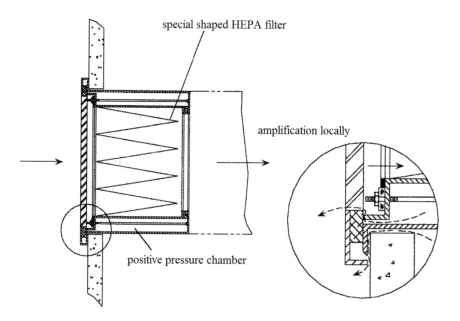

Fig. 6.5 Schematic of the negative pressure high efficiency air exhaust equipment with dynamic air current

6.2.2 *Experiment on the Device*

During the experiment, the artificial leakage point was set. The rod or circular pipe was inserted under the sealing gasket, which formed the leakage pathway. The flow rate of the supplied air was adjusted to maintain the positive pressure inside the positive pressure cavity. Test was performed to verify if the polluted air can be effectively prevented from passing through the leakage pathway to the air exhaust system.

The metal rod with the square cross section was inserted under the sealing gasket, which is shown in Fig. 6.6.

The area of the cross section for the metal rod was 2.25 mm^2. One rod was inserted at each edge. So the total number of the inserted rods was four. Two triangular gaps were formed between the side surface of the rod and the sealing gasket, whose area was equivalent with that of the cross section for the rod. The total area for the cross section of the leakage airflow was $4 \times 2.25 = 9$ mm^2. By adjusting the flow rate of the supplied air, a certain amount of positive pressure was maintained inside the positive pressure cavity. The particle concentration in the exhaust air was measured to see if it was ≥ 3 pc/L. According to Chinese standard *"Architectural and technical code for biosafety laboratories"*, there is the leakage when the scanning result is ≥ 3 pc/L.

The hard plastic tube with inner diameter 1.65 mm can be inserted under the sealing gasket, which is shown in Fig. 6.7. Four tubes in total were inserted at four edges. The total area for the cross section of the leakage airflow was 8.55 mm^2. The area of the leakage pathway at each edge of the tube is equivalent with that of the tube, which was also 8.55 mm^2.

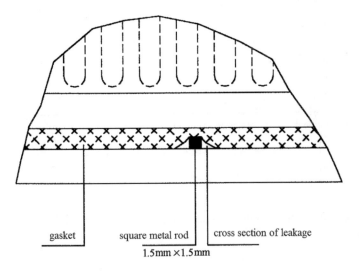

Fig. 6.6 Schematic for insertion of metal rod under the sealing gasket

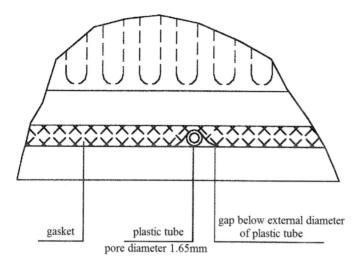

gasket   plastic tube   gap below external diameter
                                        of plastic tube
pore diameter 1.65mm

Fig. 6.7 Circular tube inserted under the sealing gasket

With the aforementioned same method, by adjusting the flow rate of the supplied air, a certain amount of positive pressure was maintained inside the positive pressure cavity. The particle concentration in the exhaust air was measured to see if it was ≥ 3 pc/L. The scanning results are shown in Tables 6.1, 6.2, 6.3, 6.4 and 6.5.

Based on the above experimental results, the following conclusions could be obtained:

(1) When the positive pressure ≥ 1 Pa relative to the room was kept in the positive pressure cavity, dynamic airflow will be forced outside of the cavity through the gap, which could prevent the hazardous air from sucking into the gap to be exhausted out.

(2) As long as the positive pressure ≥ 1 Pa relative to the room was kept in the positive pressure cavity, when the gland nut was unfastened completed or the loosening distance for the gland nut was 0.83 mm, or when the leakage area of the artificial leakage pathway was 9–17.1 mm², it has no influence on the flow rate of the leakage air. No matter whether the background concentration was subtracted, which varied from 5000 to 85000 pc/L, the concentration at the leakage point was zero or below the background concentration. Meanwhile, it was found that the phenomenon of leakage can be prevented as long as the positive pressure 1 Pa was kept inside the positive pressure cavity.

(3) After the positive pressure inside the cavity reached 1 Pa, the increase of the positive pressure has no influence on the performance for preventing the leakage airflow. This means that the flow rate of the non-hazardous air into the positive pressure cavity could be very small. The flow rate of the dynamic air through the gap is also very small.

Table 6.1 Result one by unfastening the nut

Particle concentration before test: 35000 pc/L (≥0.5 μm)

No.	Flow rate of supplied air, m³/h	Pressure value in the positive pressure cavity, Pa	Loosening distance for the gland nut, mm	Concentration (≥0.5 μm) at the leakage point near the frame at the air outlet surface, pc/L	Concentration without the influence of the background concentration, pc/L
1	2.4	+1	0.83	2	
2	2.4	+1	0.83	2	
3	2.4	+1	0.83	2	
4	2.4	+1	0.83	2	
5	2.4	+1	0.83	3	
Avg.				2.2	0.3
6	3.6	+6	0.83	3	
7	3.6	+6	0.83	2	
8	3.6	+6	0.83	3	
9	3.6	+6	0.83	2	
10	3.6	+6	0.83	1	
Avg.				2.2	0.3

Particle concentration before test: 35000 pc/L (≥0.5 μm)

No.	Flow rate of supplied air, m³/h	Pressure value in the positive pressure cavity, Pa	Loosening distance for the gland nut, mm	Concentration (≥0.5 μm) at the leakage point near the frame at the air outlet surface, pc/L	Concentration without the influence of the background concentration, pc/L
11	4	+18	0.83	0	
12	4	+18	0.83	3	
13	4	+18	0.83	1	
14	4	+18	0.83	3	
15	4	+18	0.83	2	
Avg.				1.8	~0
16	4.6	+30	0.83	1	
17	4.6	+30	0.83	1	
18	4.6	+30	0.83	2	
19	4.6	+30	0.83	3	
20	4.6	+30	0.83	1	
Avg.				1.6	~0

Note Since average concentration was used to describe the background value, negative value may appear with subtraction, which was impossible. Therefore, for those with negative value, zero was used

Table 6.2 Result two by unfastening the nut

Particle concentration before test: 15000 pc/L (≥0.5 μm)

No.	Flow rate of supplied air, m³/h	Pressure value in the positive pressure cavity, Pa	Loosening distance for the gland nut, mm	Concentration (≥0.5 μm) at the leakage point near the frame at the air outlet surface, pc/L	Concentration without the influence of the background concentration, pc/L
1	0	−7	1.33	315	
2	2.4	+1	1.33	0	
3	2.4	+1	1.33	3	
4	2.4	+1	1.33	2	
5	2.4	+1	1.33	1	
6	2.4	+1	1.33	2	
Avg.				1.6	~0
7	3.6	+2	1.33	3	
8	3.6	+2	1.33	1	
9	3.6	+2	1.33	0	
10	3.6	+2	1.5	1	
11	3.6	+2	1.5	2	
Avg.	4.6			1.4	~0
12	4.6	+1	Unfastened	2	
13	4.6	+1	Unfastened	1	
14	4.6	+1	Unfastened	3	
15	4.6	+1	Unfastened	1	
16	4.6	+1	Unfastened	2	
Avg.				1.8	
17	0	−10	Unfastened	400	

Table 6.3 Result by inserting the metal rod

Particle concentration before test: 37000 pc/L (≥ 0.5 μm)

No.	Flow rate of supplied air, m³/h	Pressure value in the positive pressure cavity, Pa	Loosening distance for the gland nut, mm	Area of the leakage pathway near the metal rod, mm²	Concentration (≥ 0.5 μm) at the leakage point near the frame at the air outlet surface, pc/L	Concentration without the influence of the background concentration, pc/L
1	2.4	+1	1.17	9	2	
2	2.4	+1	1.17	9	3	
3	2.4	+1	1.17	9	3	
4	2.4	+1	1.17	9	1	
5	2.4	+1	1.17	9	1	
6	2.4	+1	1.17	9	1	
7	2.4	+1	1.17	9	0	
8	2.4	+1	1.17	9	0	
9	2.4	+1	1.17	9	1	
10	2.4	+1	1.17	9	1	
Avg.					1.3	~0
11	3.6	+1	1.33	9	1	
12	3.6	+1	1.33	9	1	
13	3.6	+1	1.33	9	3	
14	3.6	+1	1.33	9	2	

Particle concentration before test: 37000 pc/L (≥ 0.5 μm)

No.	Flow rate of supplied air, m³/h	Pressure value in the positive pressure cavity, Pa	Loosening distance for the gland nut, mm	Area of the leakage pathway near the metal rod, mm²	Concentration (≥ 0.5 μm) at the leakage point near the frame at the air outlet surface, pc/L	Concentration without the influence of the background concentration, pc/L
15	3.6	+1	1.33	9	3	
16	3.6	+1	1.33	9	0	~0
17	3.6	+1	1.33	9	3	
18	3.6	+1	1.33	9	2	
19	3.6	+1	1.33	9	0	
20	3.6	+1	1.33	9	2	
21	4.6	+1	Unfastened	9	2	
22	4.6	+1	Unfastened	9	0	
23	4.6	+1	Unfastened	9	1	
24	4.6	+1	Unfastened	9	1	
25	4.6	+1	Unfastened	9	1	
26	4.6	+1	Unfastened	9	1	
Avg.					1.4	~0
27	0	−9	Unfastened	9	900	

Table 6.4 Result one by inserting the vent pipe

Particle concentration before test	58000 pc/L (≥0.5 μm)						Particle concentration before test	58000 pc/L (≥0.5 μm)					
No.	Flow rate of supplied air, m³/h	Pressure value in the positive pressure cavity, Pa	Loosening distance for the gland nut, mm	Area of the leakage pathway in vent pipe, mm²	Concentration (≥0.5 μm) at the leakage point near the air frame outlet surface, pc/L	Concentration without the influence of the background concentration, pc/L	No.	Flow rate of supplied air, m³/h	Pressure value in the positive pressure cavity, Pa	Loosening distance for the gland nut, mm	Area of the leakage pathway in vent pipe, mm²	Concentration (≥0.5 μm) at the leakage point near the air frame outlet surface, pc/L	Concentration without the influence of the background concentration, pc/L
1	4.6	+1	Unfastened	17.1	0		6	4.6	+1	Unfastened	17.1	2	
2	4.6	+1	Unfastened	17.1	0		7	4.6	+1	Unfastened	17.1	2	
3	4.6	+1	Unfastened	17.1	2		Avg.				17.1	1.3	~0
4	4.6	+1	Unfastened	17.1	1		8	0	−7	Unfastened	17.1	1500	
5	4.6	+1	Unfastened	17.1	2								

Table 6.5 Result two by inserting the vent pipe

Particle concentration before test	83000 pc/L (≥0.5 μm)					
No.	Flow rate of supplied air, m³/h	Pressure value in the positive pressure cavity, Pa	Loosening distance for the gland nut, mm	Area of the leakage pathway in vent pipe, mm²	Concentration (≥0.5 μm) at the leakage point near the frame at the air outlet surface, pc/L	Concentration without the influence of the background concentration, pc/L
1	4.6	+1	Unfastened	17.1	1	
2	4.6	+1	Unfastened	17.1	2	
3	4.6	+1	Unfastened	17.1	1	
4	4.6	+1	Unfastened	17.1	2	
5	4.6	+1	Unfastened	17.1	2	
6	4.6	+1	Unfastened	17.1	1	
Avg.					1.5	~0
7	4.6	+1	Unfastened	17.1	2900	

6.2.3   Operation Method

Figure 6.8 shows the shape of the negative pressure high efficiency air exhaust equipment with dynamic air current.

Before this equipment is installed at the air return (or exhaust) opening, HEPA filter is pulled out for the leakage test on the testing vehicle as shown in Fig. 6.9. The method and the standard for in situ scanning leakage test have been introduced in Sect. 6.1.

Table 6.6 presents the specification of this equipment.

After air filter is determined to be leakage-free, it is installed inside the above air exhaust device. Then the device is installed at the air exhaust (or return) opening.

Fig. 6.8 Shape of the air return (exhaust) device with leakage-free air filter

Fig. 6.9 In-situ leakage testing vehicle

Table 6.6 Specification of the air return (exhaust) device with leakage-free air filter

No.	Type	Flow rate, m³/h	Configuration specification (W × H × D)	Specification of air filter, mm	Specification of air exhaust opening, mm	Size of opening hole, mm
1	WLP-1	300	506 × 406 × 350	400 × 300 × 120	250 × 120	450 × 350
2	WLP-2	500	606 × 456 × 350	500 × 350 × 120	400 × 120	550 × 400
3	WLP-3	700	706 × 506 × 380	600 × 400 × 120	500 × 130	650 × 450
4	WLP-4	900	806 × 556 × 380	700 × 450 × 120	600 × 140	750 × 500

With the sealing by dynamic airflow, we can know that the frame is leakage-free as long as the manometer value monitored is positive "+", which is connected to the positive pressure cavity. According to GB5059 "*Code for construction and acceptance of cleanroom*", the positive pressure value should not be smaller than 10 Pa. This is much larger than the aforementioned value 1 Pa. So it is no doubt that there is no leakage.

Before we replace the air filter, the disinfection agent should be used on the front return air grille. The membrane with non-drying glue strake on the produce itself should be pasted on the edges of the air filter. Then the air filter is removed, and placed into the plastic bag.

6.3 Theoretical Analysis for Sealing with Dynamic Air Current [2]

6.3.1 Physical Model

Air exhaust system is turned on in the room at first, and then the air supply system is turned on. This period is very short. When the steady state indoors is reached, the relationships between various parameters are as follows.

(1) When the steady state is reached, the absolute pressure indoors reaches −60 Pa (−60 Pa is a very important requirement for Class Three biosafety laboratory. For the isolation ward, this value is an extremely high standard).

(2) The type of HEPA filter for supplied air is A-type. When it is operated under 80% of the rated flow at most, the pressure in front of the air filter should be 92 Pa at least.

(3) The type of HEPA filter for exhaust air is B-type. When it is operated under 80% of the rated flow at most, the pressure with the sucking force should be 236 Pa.

(4) The maximum pressure in the positive pressure cavity should be 30 Pa.

(5) Artificial gap is made on the upper sealing gasket, whose area is 17.1 mm² as labeled in the above table. The nut on the upper sealing gasket is not fastened closely. We assume the dimension of the gap is 0.1 mm (height, this is the

main sealing surface) $\times 1.6$ m (length). The nut on the lower sealing gasket is not fastened closely. We assume the dimension of the gap is 0.2 mm (height, this is the auxiliary sealing surface) $\times 1.6$ m (length).

6.3.2 Calculation

(1) Air velocity through the gap can be calculated as follows

$$v = \varphi \sqrt{\frac{2 \times \Delta P}{\rho}}$$

where ΔP is the pressure drop across the gap, Pa; ρ is the air density, which is usually 1.2 kg/m^3; φ is the air velocity coefficient, which is 0.82 in theory. But experiment shows it could be as small as 0.3–0.5. So here it is set 0.4 [2]. The air velocity through the gap on the upper sealing gasket is:

$$v_{upper} = 0.4 \times \sqrt{\frac{2 \times [30 - (-236)]}{1.2}} = 8.4 \, \text{m/s}$$

The flow rate of dynamic air current through the upper gap is:

$$Q_{upper} = 8.4 \times (17.1 + 1600 \times 0.1) \times 10^{-6} \times 3600 = 5.4 \, \text{m}^3/\text{h}$$

The air velocity through the gap on the lower sealing gasket is:

$$v_{lower} = 0.4 \times \sqrt{\frac{2 \times [30 - (-60)]}{1.2}} = 4.9 \, \text{m/s}$$

The flow rate of dynamic air current through the upper gap is:

$$Q_{lower} = 4.9 \times 1600 \times 0.2 \times 10^{-6} \times 3600 = 6.6 \, \text{m}^3/\text{h}$$

So the total flow rate through the gap is 12 m^3/h.

Since the flow rate of the supplied air indoors could reach more than 300 m^3/h, it is realistic to generate the flow rate with this value.

(2) For the extreme condition when the relative positive pressure between the cavity and the room is only 1 Pa (this is the above experimental condition), we obtain

    (a) The pressure indoors is 0 Pa, and the pressure inside the positive pressure cavity is +1 Pa. The pressure drop across the gap at the lower sealing gasket is $\Delta P = 1$ Pa.

(b) The maximum pressure drop across the gap at the upper sealing gasket will not be larger than 196 Pa, because the flow rate of the exhaust air is smaller than 80% of the rated flow.

(c) $v_{upper} = 0.4 \times \sqrt{\frac{2 \times [1-(-196)]}{1.2}} = 7.2\,\text{m/s}$

The flow rate of dynamic air current through the upper gap is:

$$Q_{upper} = 7.2 \times (17.1 + 1600 \times 0.1) \times 10^{-6} \times 3600 = 4.6\,\text{m}^3/\text{h}$$

(d) $v_{lower} = 0.4 \times \sqrt{\frac{2 \times 1}{1.2}} = 0.52\,\text{m/s}$

The flow rate of dynamic air current through the upper gap is:

$$Q_{lower} = 0.52 \times 1600 \times 0.2 \times 10^{-6} \times 3600 = 0.6\,\text{m}^3/\text{h}$$

(e) The total flow rate through the gap is 5.2 m³/h.
(f) When the pressure inside the positive pressure cavity is maintained as 1 Pa, the measured flow rate by positive pressure through the unfastened sealing strake reached 4.6 m³/h, which is close to the above calculated total flow rate. This means the assumption for the size of the gap is appropriate.

(3) Suppose the positive pressure cavity is connected with the double wall, instead of the source of positive pressure, zero-pressure is kept (the concept of zero-pressure is one patent of the author). The possible consequence will be analyzed as follows.

(a) Since the double wall or the ceiling is very air-tight, and the volume is not very large, the flow rate of the supplementary air from the air source is limited. When there is no supplementary air into the positive pressure cavity, after a certain period of leakage, the pressure inside the cavity should approach the indoor pressure. When the average pressure (0 + 60)/2 = 30 Pa is used for calculation, the air velocity through the lower gap which is connected with the room is:

$$v_{lower} = 0.4 \times \sqrt{\frac{2 \times 30}{1.2}} = 2.82\,\text{m/s}$$

(b) $v_{upper} = 0.4 \times \sqrt{\frac{2 \times (196-30)}{1.2}} = 6.7$

(c) Then we obtain: $Q_{upper} = 4.3$ m³/h and $Q_{lower} = 3.2$ m³/h.
(d) The total flow rate through the gap is 7.5 m³/h.
(e) The volume of the ceiling is assumed 25 m² × 2 m = 50 m³.
(f) Within several hours, the pressure inside the cavity will decrease much, which reaches the equilibrium state with the room. In this case, the air inside the cavity will not be forced from the cavity into the room. When there is leakage on the upper gap, the dynamic airflow cannot be formed in

the positive pressure cavity which is not connected with the air source. So the hazardous indoor air cannot be prevented from sucking through this cavity and then exhausted away. So the performance of the sealing with dynamic air cannot be established under this situation.

(g) When the ceiling is not air-tight, which is connected with the ambient by the gap, the outdoor air will be introduced through the positive pressure cavity and then into the room. Although the flow rate is not large, it cannot meet the requirement. Since the cavity is connected with the ambient, when the indoor pressure becomes positive instantaneously, indoor air may be forced into the ceiling or even the ambient, which is not permitted. Moreover, when the size of the gap on the upper sealing gasket is large and when the pressure head of the exhaust air pump is large, the pressure inside the zero-pressure cavity must be kept negative by air exhaust from the cavity, so that air could be implemented from the ceiling. When the negative pressure indoors is relatively small and when the resistance of the connecting pipe with the ceiling is large, the pressure inside the zero-pressure cavity (which is the original positive pressure cavity) will be less than the indoor pressure. In this case, the sealing performance of the zero-pressure sealing device will vanish.

(4) The following conclusions can be obtained through the above analysis:

(a) Be sure that high pathogenic aerosol in the exhaust air or the circulation air from the isolation ward must be filtered through air filter without leakage. So the air exhaust or return device must be verified that it is leakage-free.

(b) By using the positive pressure cavity with dynamic flow, the negative pressure air exhaust device can overcome the suction effect of the exhaust air with hundreds of Pascal. The leakage can be prevented. The device is leakage-free.

(c) As long as the pressure between the positive pressure cavity and the room is maintained 1 Pa, the leakage-free can be realized.

(d) Even when this device is applied, the common requirement for the closely fastening of the air filter frame should be satisfied. The sealing quality of the air filter and its installation cannot be ignored once this device is utilized.

References

1. Z. Xu, *Fundamentals of Air Cleaning Technology and Its Application in Cleanrooms* (Springer Press, Berlin, 2014), p. 636
2. Z. Xu, Y. Zhang, Q. Wang, F. Wen, Y. Zhang, R. Wang, H. Liu, W. Niu, J. Shen, High efficient negative exhausting device with dynamic ariflow sealing, the 18th ICCCS International Sympothium on Contamination Control, Beijing, 656–658 (2006)

Chapter 7
Design Points for Negative Pressure Isolation Ward

7.1 Classification

7.1.1 Classification of Infectious Diseases

According to *"Law of the People's Republic of China on the Prevention and Control of Infectious Diseases"* issued on December 1st in 2004, infectious diseases can be classified as the first class, the second class and the third class.

Infectious diseases in the first class include the plague and the cholera.

Infectious diseases in the second class include the severe acute respiratory syndrome (SARS), the human immunodeficiency virus (HIV), the viral hepatitis, the poliovirus, the human avian influenza, the measles, the Epidemic Hemorrhagic Fever, the rabies virus, the Japanese Encephalitis Virus, the dengue virus, the colletotrichum gloeosporioides, the bacillary and amoebic dysentery, the pulmonary tuberculosis, the typhoid and paratyphoid fever, the epidemic cerebrospinal meningitis, the pertussis, the diphtheria, the neonatal tetanus, the scarlet fever, the brucellosis, the gonorrhea, the syphilis, the leptospirosis, the schistosomiasis, and the malaria.

Infectious diseases in the third class include the influenza, the epidemic mumps, the rubella, the Acute Hemorrhagic Conjunctivitis, the leprosy, the epidemic and local typhus, the leishmaniasis, the echinococcosis, the filariasis, and the infectious diarrhea which is not caused by the cholera, the bacillary and amoebic dysentery, and the typhoid and paratyphoid fever.

It is also specified in this law that the preventive control measures with the first-class infectious diseases should be applied for those infectious diseases in the second class, including SARS, the pulmonary anthrax and the human avian influenza. For the infectious diseases in the second class and those with unknown origin, which need the preventive control measures with the first-class infectious

© Springer Nature Singapore Pte Ltd. 2017
Z. Xu and B. Zhou, *Dynamic Isolation Technologies in Negative Pressure Isolation Wards*, DOI 10.1007/978-981-10-2923-3_7

diseases, it must be approved, published and implemented by the state council after the epidemic incidence is reported immediately by the health administrative department of the state council.

Now more attention has been paid on the Ebola virus.

7.1.2 Classification of Isolation Wards

1. Isolation ward can be classified as the infectious isolation ward and the protective isolation ward (the isolation ward for curing the mental disease is not included in this book).

   The infectious isolation ward is also termed as the negative pressure isolation ward. It is mainly used for prevention of the airborne disease from infecting both the environment outside the ward and the people except for the patient. These diseases include the tuberculosis, the chickenpox, the pneumonia, SARS, the hemorrhagic fever virus, etc.

   This book focuses on the negative pressure isolation ward only.

   Some infected patients may also need protection, such as tuberculosis patients. In occasions when there is no infectious patient inside, the isolation ward can be used as the ordinary ward. This has been clearly specified in the standard of some nations, such as the AIA standard in U.S.A. "When there is no need for isolation, the isolation ward can be used as the ordinary nursing room or can be divided into individual isolation wards." "When there is not patients with airborne diseases, the isolation ward can be used for patients without infectious diseases" [1].

2. The infectious strength of the disease on patient inside the infectious isolation ward can be classified as four levels which can be referred in Chap. 1.

   The function of the isolation ward is as follows:

   (1) The isolation ward should play the role of isolation.
       Isolate the ward from the ambient environment, and isolate the patient from the medical personnel.
   (2) The isolation ward should play the role of safety.
       To guarantee the safety of the environment outside the ward.
       To guarantee the safety of the medical personnel inside the ward.
       It could also be classified as four levels according to the required pressure difference, such as the Australian standard which is shown in Table 7.1 [2].

Table 7.1 Classification of isolation ward in "Guidelines for the classification and design of isolation rooms in health care facilities" from Australia

Classification	S level (zero pressure)	N level (negative pressure)	A level (alternative pressure)	P level (positive pressure)
Ventilation	No pressure difference between the ward and the adjacent corridor	Pressure inside the ward is lower than that in the corridor	Positive or negative pressure can be switched inside the ward with the ventilation facility	Pressure inside the ward is larger than that in the corridor
Method of infectious prevention	Prevent the contact infection and airborne infection	Prevent the airborne infection	Not recommended	Prevent the exterior pathogen in the ambient environment from infecting the HIV infected patient
Indoor hand washing	Yes	Yes		Yes
Bathroom (shower bath, closestool, hand-washing)	Yes	Yes		Yes
Door with switch	Yes	Yes		Yes
Air-lock room (buffer room or antechamber)		Yes	Not recommended	Optional
Air-tight Room, grille above door to control airflow direction		Yes		Yes
Example	Patient infected with the vancomycin-resistant enterococci, the gastroenteritis, the windburn, the hepatitis A, and the neisseria meningitides	Patient infected with the measles, the chickenpox, suspected or confirmed pulmonary tuberculosis or laryngeal tuberculosis	Not recommended	Prevent the patient after the marrow transplantation from infected by aspergillus

7.2 Layout Plan

7.2.1 Environment

1. The distance between the isolation ward and the surrounding buildings especially the dormitories and the public buildings should be 20 m at least [3].
   This has been pointed out for the site selection of the infectious diseases hospital in the national standard GB50849-2014 *"Code for design of infectious diseases hospital"*. Since the isolation ward is the main component of hospital, this principle should be followed naturally. The concept of the minimum distance 20 m was first proposed by author, which was adopted by the national standard GB 50346-2004 *"Architectural and technical code for biosafety laboratories"*. This is aimed in terms of the risk of the exhaust air from the biosafety laboratories. When the microbes with highly pathogenic and the safety of the exhaust air are concerned, the isolation ward should not be an exceptional.
2. In the general hospital, it is better if the isolation ward can be placed alone. Otherwise it should be set at one edge of the building as possible. If should form a zone and placed at the leeward wind side in the most annual wind direction in this area. When there are two most wind directions in this area, the isolation ward should be placed at the opposite direction facing the wind direction with the least frequency.

7.2.2 Partition

1. For the ward are consisting of multiple isolation wards, the clean area, the potentially polluted area and the polluted area should be strictly distinguished. In general, the ward itself (including its bathroom) and the area for activity of patients are the polluted area. The public corridor is the potentially polluted area. The primary changing room, the preparatory room and the offices for the medical personnel area the clean area.
2. If there is enough space for setting double corridors, the patient should enter into the ward through the back corridor and the back door. In this case, the back corridor should be classified as the polluted area. The relative pressure inside the back corridor is positive. The net width of the corridor should be not less than 2.4 m. When there is relative height difference in the corridor, the barrier-free cohesion edge should be used, and the antiskid measures should be taken.
   The medical personnel should enter into the ward through the front corridor which should be classified as the potentially polluted area. Only the region outside the exit of this corridor could be classified as the clean area.

Figure 7.1 shows the renovated SARS ward in a hospital in Shanghai, where the scheme with two corridors was adopted [2]. In the figure, the semi-clean area can be considered as the potentially polluted area. While the semi-polluted area can also be considered as the polluted area, where the pressure is slightly higher than that in the ward. Besides, buffer rooms should be set at two edges of this corridor.

As shown in Fig. 7.1, it is better if there is also the buffer room at the back door of the ward. In this case, it is unnecessary to set the buffer room at two edges of the exterior corridor which can be classified as the potentially polluted area.

3. In the national standard GB50849-2014 "*Code for design of infectious diseases hospital*", more than two passageways should be set for the ward in the hospital where the number of the sickbeds is more than 150, and two passageways are required to be set for the ward in the hospital where the number of the sickbeds is less than 150. The detailed information is given in Table 7.2 (Both Tables 7.2 and 7.3 are cited from the literature "*Discussion on planning design of the infectious diseases hospital*" by Zhang Chun-yang and Huang Kai-xin). The items 2 and 3 in the table are united.

If the ward is placed in other buildings, the passageway must be set independently, which could be disinfected in a closed space.

4. According to the different transmission routes of the infectious diseases, the ward and the outpatient room (or the emergency room) in the hospital are set. The ward with infectious disease by airborne transmission must be set independently, which is shown in Table 7.3.

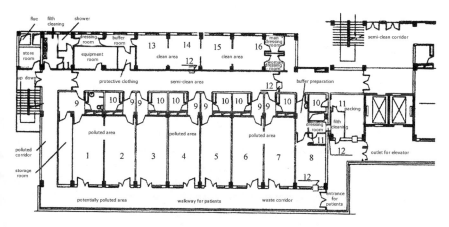

Fig. 7.1 Layout of the renovated SARS ward. *1–7* Ward; *8* Analysis laboratory; *9* Buffer room; *10* Bathroom; *11* Disinfection room; *12* Delivery window; *13–15* Offices for the director, the doctor and the nurse, respectively; *16* Treatment room

Table 7.2 Setting of streamline and general passageway in the infectious diseases hospital

Number of passageway	Property	Streamline inside hospital	Location of passageway	Key point
Three passageways	Clean area	(1) Medical personnel, discharged patient, visitor, cleaning suppliers (food and medicine, etc.)	The primary road	(1) The streamlines are clearly set, which could share the general passageway
	Polluted area	(2) Contaminated goods (corpse and waste, etc.)	The secondary road	(2) The passageway should be set far away from the living area. It is better to be close to the morgue. The door is locked during the normal times
		(3) Patient	The secondary road in the city with convenient transportation	(3) It is set inside the isolated region, which is the pathway for the infected patient and the ambulance. If it must be set at the same side with the passageway for the cleaning suppliers because of the space limit, the distance between them must be clearly specified for avoiding confusion

7.3 Isolation Ward

7.3.1 Ward

1. The isolation wards can be classified as single room, double room and multiple room. In the local standard DB 11/663-2009 "*Essential construction requirements of negative pressure isolation ward*" was issued in Beijing (which is shortened as "*Essential requirements*"), it is required that the maximum number of occupants is three. While in standards from Germany and U.S.A., it is required that the maximum number of occupants in each ward is two, and for the renovated ward the corresponding value is four.

Table 7.4 shows the data on the area of the ward by "*Essential requirements*" and standard from U.S.A.

In hospitals from the Netherlands, the width of the sickbed is 1 m, and the minimum distance between sickbeds is 1.5 m. In U.S.A., the distance between sickbed is required 2.24 m.

Table 7.3 Setting of the ward and the outpatient room (or the emergency room)

Name	Feature of diseases	Keypoint for design of building
Infectious ward for digestive tract	When the pathogen (such as typhoid fever, bacillary dysentery, cholera) contaminates the food, the drinking water or the dishes, the susceptible is infected during intaking of food	Isolation measures were taken according to the feature of the disease. Individual bathroom must be set inside every ward (or treatment room). The waste is disinfected and sterilized, and then discharged into the drainage system of the hospital
Infectious ward for respiratory tract	For the pathogen such as measles, diphtheria and tuberculosis, they are likely to be inhaled by the susceptible once they are suspended in the air, which will cause the infection	Clean area, potentially polluted area and polluted area are strictly distinguished. Organized air distribution is required for the airflow in the ward. The pressure in the clean area should be larger than that in the potentially polluted area. The pressure of the potentially polluted area should be higher than that in the polluted area. The negative pressure is kept inside the ward, so that the polluted air cannot disperse towards outside. It is better that the treatment room should be set independently, which is located at the bottom
Infectious ward (or treatment room) for entomophily	When bloodsucking arthropods such as mosquito, human louse, rat flea, sandfly and chigger mite bite the susceptible, the infection by the pathogen will occur. Malaria, epidemic typhoid fever, endemic typhoid fever, kala-azar and tsutsugamushi disease will be induced respectively	Attention should be paid on killing the mosquito and defaunate in the ward (or treatment room). Insect screen should be installed at the door and the window locations. Blacklight trap can be used to kill the mosquito and the fly at the window location, which can prevent the disease transmission through the mosquito and the fly
Infectious ward (or treatment room) for other diseases (blood source)	Pathogens such as malaria, Hepatitis B, Hepatitis C and HIV virus exist in the blood or the body fluid of the carrier or the patient. Transmission occurs during the application of the blood product, the delivery process or the sexual intercourse process	The hand-washing apparatus should be set in the passageway of the medical personnel for each ward (or treatment room). Responsive switch should be used. Disinfection and hand-washing should be carried out when the medical personnel walk in and out of the ward (or treatment room)
Special ward (or treatment room)	It is aimed to treat the sudden and the fulminating infectious disease such as SARS	It should be set independently. Clean area, potentially polluted area and polluted area are strictly distinguished. Buffer room should be placed between different areas. Organized air distribution is required for the airflow in the ward. The negative pressure is kept inside the ward. The polluted air can be discharged outside after it is disinfected

Table 7.4 Area of the isolation ward (the bathroom is not included)

	Standard value	Minimum value	Minimum distance between sickbed and any obstacle	Standard value	Minimum value	Minimum distance between sickbeds
U.S. A.	11.2 m^2	9.3 m^2	0.91 m	9.3 m^2	7.5 m^2	2.24 m^2
China	11 m^2 (net value)	9 m^2 (net value)	0.9 m	9 m^2 (net value)	7.5 m^2 (net value)	1.1 m

In short, the space inside the ward should be spacious enough to place equipments such as the bedside X-ray machine and the breathing machine. Therefore, the values of the distances are larger than that in the ordinary ward.

According to "*Essential requirements*", the net height of the ward should not be less than 2.8 m.

7.3.2 Accessory Rooms

1. Except for the isolation, the intensive care unit, the office for doctor, the office for nurse, the nurse station, the disposal room, the treatment room, the duty room, the base for bed and cloth, the canteen preparation room and the room for boiling water should also be placed inside the ward area. When the number of the wards is large, X-ray room should be set. When there is the teaching task, the demonstration classroom should be set.
2. The bathroom should be annexed in the isolation ward, which includes the closet bowel flushed with sterilized water, the shower, the washbasin with inductive tap. Door should not be placed for other bathrooms outside of the ward. Instead, the open labyrinth-style inlet can be adopted.
3. The canteen preparation room in each ward area should distinguish the clean room from the polluted room. The delivery window is set between them. If the disposable tableware is used, the canteen preparation room can be set inside the clean area, and no isolation is needed.
4. If needed, biosafety cabinet should be set for testing of special specimens, which is operated inside the Class 2 biosafety laboratory.
5. The autopsy room for the infected patients especially those with the severe acute respiratory syndrome (SARS) should be designed according to Class 3 biosafety laboratory. According to the Japanese standard, the anatomy station should be placed inside the unidirectional flow unit.

7.3.3 People Flow and Goods Flow

1. Delivery window should be placed on the wall of the corridor adjacent to the isolation ward, which is used for deliver the medicine and the food. According to the nation standard JG/T 382-2012 "*Pass box*", the classification of the delivery window is shown in Table 7.5. The delivery window set between the isolation ward and the polluted area should be Type B1 or C1. Figure 7.2 shows the basic appearance of the delivery window.
2. Air shower should not be placed along the passage of people flow. Air curtain should not be used at the gate in the ward area.

Table 7.5 Classification of the delivery window

Type	Symbol	Function
Basic	A	It is installed on the partition wall of the ward, which is used for delivery of goods. Two doors are inter-locked. It has the basic function of isolation for air from rooms in two sides of the partition wall
With function of air cleaning	B1	Self-purification system composed of fan and HEPA filter is included, which has the function of air cleaning for aerosol inside the delivery window
	B2	It includes the air supply system and the air exhaust system containing HEPA filters, which has the function of removing aerosol inside the delivery window and aerosol exhausted out of the delivery window
	B3	It has the function of air shower. The particles on the surface of the delivered goods inside the unit are cleaned by high speed clean airflow injected through the nozzles
With function of disinfection	C1	UV-light tube is installed in the passage of the delivery window. The microbes on the surface of the wall and the delivered goods, and the air through the passage should be disinfected if needed
	C2	Inlet and outlet of the sterilization gas are set on the wall of the delivery window. The interior surface of the delivery window can be disinfected by the disinfection apparatus outside the delivery window when it is connected with the inlet and outlet of the sterilization gas if needed
Negative pressure	D	Under the operational condition, the negative pressure with certain value can be maintained inside the delivery window
Air-tight	E1	There is no visible gas leakage
	E2	When the pressure in the passage reaches -500 Pa, the decay rate of the negative pressure within 20 min is less than 250 Pa

 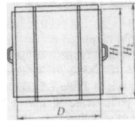 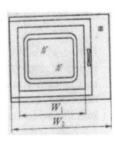

Fig. 7.2 Basic and ordinary delivery window

3. The ordinary vertical hinged door or the upper suspended sliding door can be used between the isolation ward and the buffer room. The vertical hinged door is suitable to be set between the buffer room and the corridor. In both cases, the wooden door should not be used.
4. When it is possible, the advantage is more obvious if the sliding door is adopted between the isolation ward and the buffer room. In this case, the velocity of the induced turbulence by the sliding door is the minimum, compared with that of the vertical hinged door. Therefore, the sliding door is recommended for the entrance of the isolation ward in related standards in Japan. This is also recommended in AIA standard in U.S.A. and it is also pointed out that the sliding slot should not be set on the floor for this kind of the sliding door. Therefore the upper suspected sliding door should be used, which is the same as that of the clean operating room. Of course, whether the sliding door can be set depends on the space of the buffer room. The gap of the vertical hinged door is small, so it should be used between the buffer room and the corridor, which is suggested in Japanese standard.
5. It is not necessary to use the air-tight door and the air-tight inter-lock door for the doors of the isolation ward and the buffer room. The requirement can be met when the ordinary door is used. But the wooden door should not be used.
6. Except for the safety door and the door towards the entrance hall which open outwardly, other doors should open towards the side with larger pressure.
7. Door should not be placed for other bathrooms outside of the ward. Instead, the open labyrinth-style inlet used in the airport terminal building can be adopted.
8. Buffer room is placed outside of the isolation ward. Positive pressure is maintained in the buffer room relative the isolation ward, while negative pressure or zero pressure is kept in buffer room relative to the outside of the buffer room. This kind is called Three-Room-One-Buffer, or Two-Area-One-Buffer, which is shown in Fig. 3.4.
   The isolation ward is the polluted area, while the area outside of the corridor is the clean area.

7.4 Clean Air Conditioner

7.4.1 Particularity of Cleaning Air Conditioner for Isolation Ward

1. There was objection opinion for the usage of air conditioning system in the isolation ward during the early epidemic of SARS in China. Because of the incognizance of SARS, for the purpose of the emergency response, it was emphasized in the "*Design Highlights for Hospital Buildings Receiving SARS Infected Patients*" issued by the Ministry of Health of the People's Republic of China that the ventilation condition is accessible to all the areas. Central air conditioning system is prohibited in all the areas. Air exhaust unit can be installed for the simple-built negative pressure ward.

It is understandable that these provisional measures were specified. But during the East Asia Rainy Season (or Meiyu Season) and the hot summer in the southern region where the humidity is extremely large, the propagation of pathogenic microbes indoors cannot be prevented in hospitals by natural ventilation alone. In this case, the microbial contamination may still occur. If the temperature and the humidity indoors are very high, the patient will generate heat and sweat, which will increase the bacterial generation rate. When the medical personnel wear the insulation garment, the protective cloth, the mask and eyeglasses, sweating will occur soon after the work. Sometimes febrile disease may even appear. Especially in the isolation ward for SARS, the working environment will be worsened if the problem of HVAC and environmental control is not solved. This will influence the physical and mental health of the medical personnel.

In June of 2003, the WHO representative office in China wrote to Ministry of Health of China. Different opinion was proposed. Natural ventilation by window opening is not allowed in the isolation ward for SARS. The HVAC system should be operated continuously. Exhaust air fan should be installed on the exterior window or the exterior wall, so that negative pressure is maintained indoors.

However, the "*Hospital infection control guidance for Severe Acute Respiratory Syndrome (SARS)*" issued by WHO pointed out that when there is no independent air exhaust system in HVAC system, the air conditioning system can be turned off, and ventilation can be provided by window opening. But windows are prohibited to be open towards the public space.

In fact, air conditioning system is not prohibited in related foreign standards. And circulation air can be adopted conditionally.

In "*Guidelines for Preventing the Transmission of Mycobacterium tuberculosis in Health-Care Facilities*" issued by CDC from U.S.A. in 1994, it was emphasized that for the known airborne infectious droplet nuclei, the general exhaust system can be adopted. Meanwhile, it was pointed out that it was inevitable to adopt the scheme of recirculation air. When HEPA filter is used, air can be re-circulated. Three kinds of air distribution were proposed, which would be introduced in the next section. It was also pointed out in *"The guideline of design and management for air-conditioning system of Hospital"* by Healthcare Engineering Association of Japan that the fan unit with HEPA filter can be installed in the isolation ward.

In March 1st of 2003, novel idea for central air conditioning system appeared in *"Code for hygiene of central air-conditioning ventilation system in public places"* published by ministry of health of China. It was specified that when the epidemic of airborne disease occurs in the local area, the central air conditioning system can be operated continuously only when the following requirements are satisfied. It can be used when the full fresh air scheme is used. It can also be used when the air cleaning device with disinfection function is installed and its performance is guaranteed to be effective. It can also be used when independent ventilation is accessible to each room when the HVAC system scheme by the combination of the fan coil unit and the outdoor air is adopted. However, this solution is obviously not enough. It should be pointed out that for the isolation ward with infectious disease, except for the full fresh air system, air can be circulated only when HEPA filter is installed (at the return air grille). Of course, no matter whether full fresh air system is used, exhaust air must pass through HEPA filter.

For the infectious diseases hospital and isolation ward, the above idea is also applicable.

But different cleaning air conditioner system should be set for the clean area, the potentially polluted area and the polluted area, respectively.

2. The air-circulation system is only suitable for the single ward with infectious disease. Or circulation of indoor air is adopted for the ward receiving multiple people with the same kind of disease. When heating and cooling are provided by fan coil unit, another independent or public outdoor air supply system should be set. In this case, it is suitable to switch to or adopt the full fresh air system for one or more wards.

7.4.2 Specific Requirement

1. The air change rate in the isolation ward should be 8–12 h^{-1}. The flow rate of outdoor air per person should not be less than 40 m^3/h. In other auxiliary rooms, the air change rate should be 6–10 h^{-1}.
2. Although there is no requirement on the air cleanliness level in the isolation ward, air filtration device with low pressure drop and efficiency larger than the high and medium efficiency air filter shown in Table 3.20(1), (2) must be

installed at the air supply opening. HEPA filter should be installed for the supplied air into the buffer room where the air change rate is ≥ 60 h^{-1}.

3. The safely demountable leakage-free high efficiency negative pressure air exhaust device sealed with dynamic air current can be used in the isolation ward and its bathroom.

4. The outlet of the air exhaust pipeline should be linked towards outdoors. The check valve and the rain droplet prevention measures should be used. The end of the air exhaust pipeline should be above the roof by more than 2 m. It should be far away from the air intake opening on the wall by more than 20 m, and it should be located in the downstream of the air intake opening. When the distance between them is less than 20 m, the retaining fence should be set.

5. The cleaning air conditioning system should be operated for 24 h continuously. The air velocity of the supplied air in the daytime should not be less than 0.13 m/s. While in the evening, the flow rate is under the low gear station, and the velocity of the supplied air should not be larger than 0.15 m/s. The air supply and air exhaust should be inter-locked. The air exhaust system should be turned on at first and closed later.

6. In the ordinary time, the cleaning air conditioning system can provide normal pressure status in the isolation ward, when the non-infectious patients are received. But the pressure conversion scheme is not allowed in the infectious isolation ward and the protective isolation ward. This has also been specified in the related standards in Japan and U.S.A.

7. In the isolation ward where cleaning air conditioning system and air circulation system with HEPA filter are used, indoor air cleaning device should not be used again to disturb the original air distribution.

8. Air cleaning device must be installed at the entrance of the outdoor air. According to national standard GB51039-2014 *"Code for design of general hospital"*, two-stage air filters including the coarse and fine filters should be installed at least for the entrance of the outdoor air, when the annual concentration of respirable particulate matter PM$_{10}$ in the atmosphere does not exceed 0.10 mg/m^3. When it exceeds 0.10 mg/m^3, the high and medium efficiency air filter should be added.

   In European standard EN13779, it is also specified that when the outdoor air is clean, two-stage air filters such as F5 + F7 should be set for the outdoor air pipeline.

   It is also specified in ISO 16814 that G4 + F7 should be set for outdoor air. It is better to set F5 + F8. It is very good to set F7 + F9.

9. According to GB51039-2014 *"Code for design of general hospital"*, for the return air grille in the HVAC system in the auxiliary room for isolation ward area, air filters with initial pressure drop less than 50 Pa, the one-pass penetration for microbes not larger than 10% and the one-pass arrestance for particles not larger than 5% should be installed.

7.5 Positioning Arrangement of Air Openings

7.5.1 Arrangement for Single Bed

1. The air distribution scheme with upper-supply and lower-return should be adopted in negative pressure isolation ward. The total trend of airflow should be consistent with the settlement direction of particles. The airflow inside the negative pressure ward and its ward area should be directional, which flows from the clean area towards the polluted area.
2. According to the simulated and experimental results in Chap. 4, the primary and secondary air supply outlets as shown in Fig. 7.3 should be chosen for the negative pressure isolation ward. The primary air supply outlet is set on the ceiling above where the medical personnel usually stand near the sickbed. The distance from the head of the sickbed should not be larger than 0.5 m. The length should not be less than 0.9 m. The secondary air supply outlet should be set on the ceiling above the end of the sickbed. The distance from the end of the sickbed should not be larger than 0.3 m. The length should not be less than 0.9 m.
3. The ratio of the areas between the primary and the secondary air supply outlets should be between 2:1 and 3:1. The velocity of supplied air should not be less than 0.13 m/s.

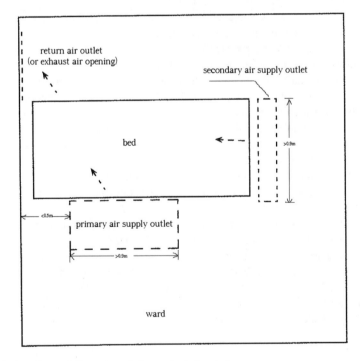

Fig. 7.3 Location and dimension of air supply outlets indoors

4. The type of double louver grille should be used for air supply outlets.
5. The type of single louver grille should be used for air return (or exhaust) opening. They should be placed at the lower side of the head of the sickbed relative to the air supply outlet. The upper edge of the air return (or exhaust) openings should be placed less than 0.6 m above the floor. The lower edge should be higher than 0.1 m above the floor. The velocity of the return air (or exhaust air) should not be larger than 1.5 m/s.

7.5.2 Arrangement for Multiple Beds

1. The layout of the air supply outlets for double beds can be chosen as shown in Figs. 7.4 and 7.5.

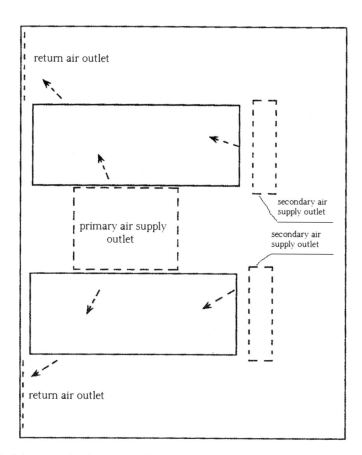

Fig. 7.4 Scheme one for air supply outlets in double-bed ward

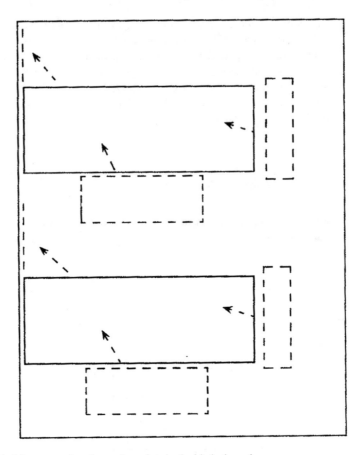

Fig. 7.5 Scheme two for air supply outlets in double-bed ward

2. The layout of the air supply outlets for multiple beds can be chosen as shown in Fig. 7.6.

   For the ward with multiple beds, it is not allowed that one bed is located at the leeward side of another ward. Figure 7.7 shows an example where infection occurred in practice. Airflow from natural ventilation or mechanical ventilation moves from one side of the room (at the window side) towards the door side (or the bathroom). This results in infection at the leeward side.

7.5.3 Arrangement of Air Openings in Buffer Room

1. When the temperature inside the ward is higher than that in the buffer room, the air return (or exhaust) openings inside the buffer room should be placed on the

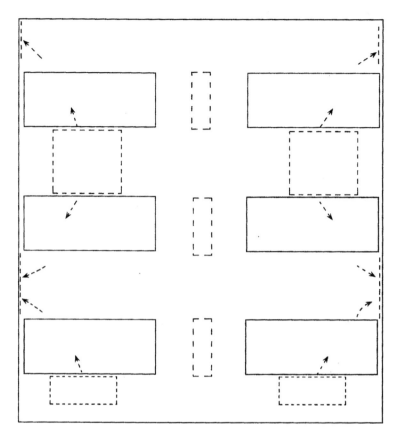

Fig. 7.6 Scheme two for air supply outlets in multiple-bed ward

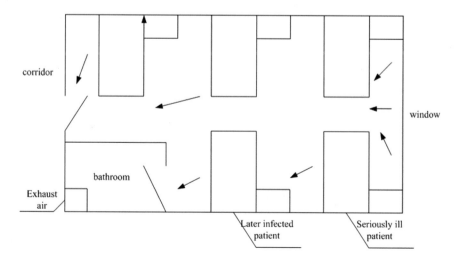

Fig. 7.7 Example of the layout of sickbeds infected at the leeward side

ceiling above the door connecting the ward. Air supply outlet should be placed on the ceiling at the opposite side relative to the air return (or exhaust) openings. They are not necessarily symmetric. Based on the analysis on the function of temperature difference in Chap. 6, when the door of the ward is open under this condition, the convection of air with air out above and air entrance down will appear. The air return opening above the door of the buffer room can remove the pollutant from the airflow leaving the ward immediately, which is shown in Fig. 7.8.

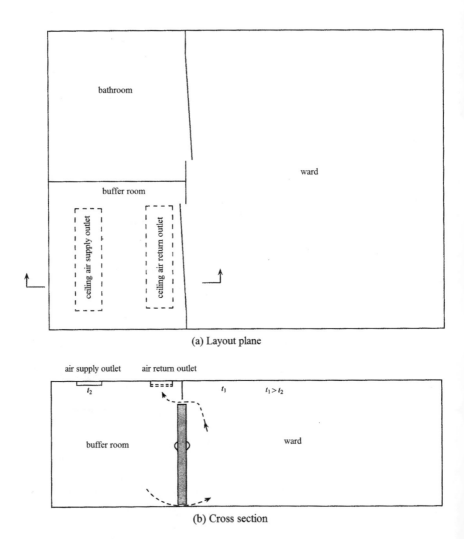

Fig. 7.8 Layout of the air openings in the buffer room (When the temperature inside the ward is higher thanthat in the buffer room, the door of the buffer room towards outside is not given in the figure.)

2. If the temperature inside the buffer room is higher than that in the ward, the air distribution with the upper-supply and lower-return scheme can be adopted inside the buffer room. This is because when the door of the ward is open, the polluted air will penetrate into the buffer room from the lower part of the door.

7.6 Determination of Pressure Difference and Differential Pressure Flow Rate

7.6.1 Pressure Difference

1. Based on the analysis on the dynamic isolation technique and DB11/663-2009 "*Essential construction requirements of negative pressure isolation wards*", the distribution of the pressure difference of the isolation wards can be set as shown in Fig. 7.8.

   The relative pressure between the ward and the buffer room, as well as that between the buffer room and the interior corridor, should not be less than 5 Pa. The sequence for the extent of negative pressure is the bathroom, the negative pressure isolation ward, the buffer room, and the interior corridor. For the ordinary negative pressure isolation ward, at least one buffer room should be set outside the ward based on the actual condition.

2. For the buffer room between the corridor inside or in front of the potentially polluted area and the clean area, its pressure relative to both this corridor and outdoor should be positive. The relative positive pressure between the buffer room and the region connecting the ambient should not be less than 10 Pa.

3. Because the ward and its bathroom are polluted areas and there is usually exhaust air system in the bathroom, the air must flow from the ward towards the bathroom. From the viewpoint of principle of dynamic isolation with dynamic current, the value of pressure difference between the ward and the bathroom was not given in DB11/663-2009. Instead, it was required for the directional airflow from the ward towards the bathroom. By adjusting the exhaust air volume, the degree of negative pressure in the bathroom should be higher than that in the ward. Upward shutter can be set above the door of the bathroom.

4. For the isolation ward with extreme high risk, two negative pressure buffer rooms can be set in series at the gate of the ward. In Japanese standard, this suggestion has been proposed. The gradient of the pressure difference is: Ward——— ← Buffer room (1)— ← Buffer room (2)— ← Interior corridor → ←Buffer room (3)(+ or −)–↔Clean area (+ or 0).

7.6.2   *Differential Pressure Flow Rate*

When it is not convenient to calculate the pressure difference or when the value of the pressure difference has not been obtained, the value of air change rate is usually used to obtain the differential pressure flow rate in engineering. It is also pointed out in CDC from U.S.A. that since the pressure difference is too small, the flow rate of the exhaust air from the room can also be used to determine the pressure difference in the room. This means when the flow rate of the exhaust air is less than 85 m^3/h, the requirement for the negative pressure is satisfied.

Based on the data in Table 2.4, For the ward where the door of the room is not air-tight, this flow rate of the exhaust air is equivalent to the pressure difference with value slightly more than 1 Pa indoors. If the pressure difference with value not less than 3 Pa as analyzed before, the corresponding flow rate of exhaust air should reach 119 m^3/h. Therefore, the suggested flow rate of exhaust air should not be less than 120 m^3/h. The corresponding flow rate of exhaust air for the pressure difference 5 Pa should be 150 m^3/h.

However, it is still a pure theoretical value with 120 m^3/h. When the differential pressure flow rate ΔQ is determined, the practical problem should also be considered.

Figure 7.9 shows the negative pressure room with supplied air and exhaust air. In the ventilation system of the room there could be fan and the adjusting valve, or fan and the constant air volume valve.

There are positive and negative deviations of the flow rate for the fan and the constant air volume valve. For example, the deviation of the TROX valve can reach ±10%, which has no relationship with the pressure. The deviation of the venturi valve can reach ±5%. In situ adjusting test shows that because the area of the ward is not large, when the ward is very air-tight, the fluctuation of extreme flow rate in the ward will influence the variation of its pressure [4].

One most unfavorable condition in the negative pressure room is that positive deviation Δq_1 appears on the supply fan or the constant volume valve. This means

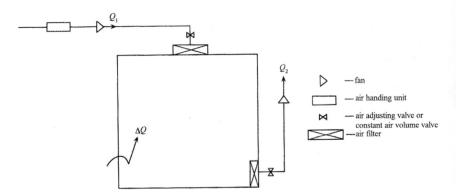

Fig. 7.9 Schematic of the negative pressure room with supplied air and exhaust air system

the actual flow rate of the supplied air is larger than the designed or rated flow. The other most unfavorable condition in the negative pressure room is that negative deviation Δq_2 appears on the exhaust fan or the constant volume valve. This means the actual flow rate of the exhaust air is smaller than the designed or rated flow. The aforementioned situations can be expressed with three conditions shown in Figs. 7.10, 7.11 and 7.12.

For condition one, we know:

$$(Q_2 - \Delta q_2) - (Q_1 + \Delta q_1) = \Delta Q' \geq \Delta Q$$

Or

$$(Q_2 - Q_1) - (\Delta q_1 + \Delta q_2) = \Delta Q' \geq \Delta Q$$

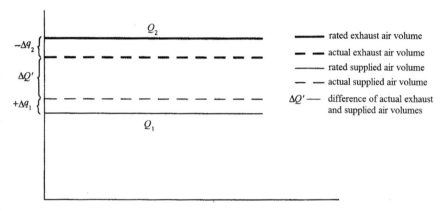

Fig. 7.10 Schematic of condition one for positive deviation for the supplied air and negative deviation for the exhaust air ($\Delta Q < \Delta Q'$)

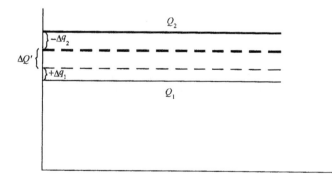

Fig. 7.11 Schematic of condition two for positive deviation for the supplied air and negative deviation for the exhaust air ($\Delta Q < \Delta Q'$)

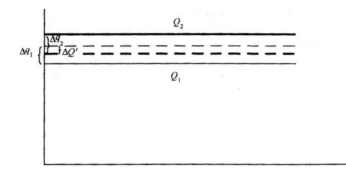

Fig. 7.12 Schematic of condition three for positive deviation for the supplied air and negative deviation for the exhaust air ($\Delta Q' < 0$)

In this case, both the flow rate and the pressure difference can be adjusted normally, which is not affected.

For condition two, we know:

$$(Q_2 - \Delta q_2) - (Q_1 + \Delta q_1) = \Delta Q' < \Delta Q$$

Or

$$(Q_2 - Q_1) - (\Delta q_1 + \Delta q_2) = \Delta Q' < \Delta Q$$

In this case, the flow rate of the exhaust air is obviously not enough. The negative pressure cannot meet the requirement, which needs further adjustment.

For condition two, we know:

$$(Q_2 - \Delta q_2) - (Q_1 + \Delta q_1) = \Delta Q' < 0$$

Or

$$(Q_2 - Q_1) - (\Delta q_1 + \Delta q_2) = \Delta Q' < 0$$

In this case, the actual flow rate of the exhaust air is smaller than that of the supplied air. Slight positive pressure appears in the ward, which needs further adjustment.

Therefore, for the negative pressure ward, the designed differential pressure flow rate of exhaust air is not only the difference of the flow rates between the exhaust air and the supplied air, but also the deviation of the fan or the adjusting valve. This means the differential pressure flow rate of the exhaust air should be larger than the summation of the leakage air flow rate and the absolute value of the positive and negative deviation for the flow rate of fan and adjusting valve. This means:

Differential pressure flow rate of exhaust air $= \Delta Q + (\Delta q_1 + \Delta q_2) > Q_2 - Q_1$

$$(7.1)$$

Now the easily appearance of the problem can be better explained for the air-tight room. When the room is too air-tight, the leakage flow rate is very small. In this case, the difference of the flow rate between the exhaust air and the supplied air is very small. If the value of this difference is smaller than the summation of the absolute value of the positive and negative deviation for the flow rate, it is difficult to adjust the pressure difference inside the room. The pressure always fluctuates, or slight positive pressure appears.

Therefore, the real differential pressure flow rate should be the summation of the above differential pressure flow rate value 120 m^3/h and the absolute value of the positive and negative deviation for the flow rates for supplied and exhaust air. In this case, it may be ≥ 150 m^3/h.

7.6.3   Expression of Pressure Difference

1. Compared with the atmospheric pressure, the relative pressure inside each room is marked in the frame as shown in Fig. 7.13.

Fig. 7.13 Expression of absolute pressure difference

2. Compared with the pressure in the adjacent room, the direction of the pressure difference is expressed with the arrow at the gate connecting two rooms. The value of the pressure difference is marked beside the arrow or at the end of the arrow. Take the above figure as an example, Fig. 7.13 can be obtained. In Fig. 7.13, the value "−10 Pa" means the pressure difference between the room where the arrow starts and the room where the arrow points to is −10 Pa. The value "+5 Pa" means the pressure difference between the room where the tail of the arrow starts and the room where the arrow points to is +5 Pa. Therefore, both the value of the pressure difference and the location cannot be confused. It is shown in Figs. 7.13 and 7.14 that these two kinds of expression methods are different.

3. In terms of the pressure difference itself, it is aimed to prevent the leakage from adjacent room or the outward leakage towards adjacent room. Because the room studied is connected with the adjacent room, the pressure difference between them is meaningful.

However, it has no practical implication to discuss the leakage towards the ambient air through the envelope or the inward leakage from the ambient air towards indoors. The biosafety laboratory with high class level is an exceptional.

In the layout shown in Fig. 7.15, it is required that:

The absolute value of negative pressure in room A $\Delta P_{AO} \nleq 40$ Pa.

The absolute value of negative pressure in corridor B $\Delta P_{BO} \nleq 30$ Pa.

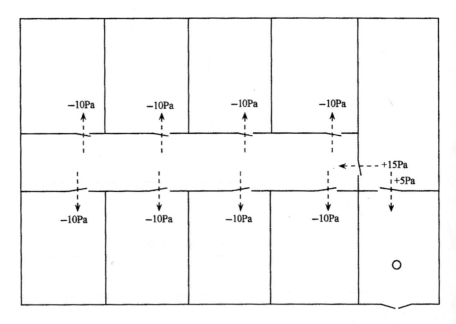

Fig. 7.14 Expression of relative pressure difference

Fig. 7.15 Absolute pressure
difference in the plane design

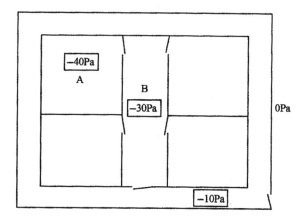

If all the measured data are −40 Pa, the pressure in corridor B does not violate
the original requirement. But in fact there is no relative pressure difference
between A and B. The design or adjusting results are not reasonable.

Figure 7.16 shows another example. The differential pressure gauge only set
on the partition wall between the isolation ward and the polluted corridor,
which shows the relative pressure difference between them. Inspection shows
that there is no differential pressure gauge between the isolation ward and the
buffer room, but the actual pressure difference is 0. In this case, even though
the pressure difference between the isolation ward and the polluted corridor
meets the requirement, there is no pressure difference between the isolation
ward and the buffer room. The buffer room and the isolation ward is combined
to be one unit. The function of buffer loses, so the buffer room only exists as a
name.

4. It has been mentioned before that it is difficult to control with the absolute
   pressure during the actual adjusting process. If one side of the differential
   pressure sensor connects indoors while the other connects outdoors as shown in
   Fig. 7.17, the sampled pressure difference is unstable because of the wind
   direction and wind magnitude of the atmosphere. It may be positive or negative,
   large or small.

   In fact, during the process of system design, the flow rate by negative pressure
   or positive pressure is calculated based on the relative pressure difference.
   Figure 7.13 should be expressed as Fig. 7.18. The difference of the exhaust air
   volume from A and that from B can be calculated. The difference of the exhaust
   air volume from B and that from the exterior corridor can also be calculated.
   This has been introduced in Sect. 2.1.2.

   Based on the relative pressure, the corresponding absolute pressure difference
   relative to atmospheric pressure can be converted. The conversion equation is as
   follows:

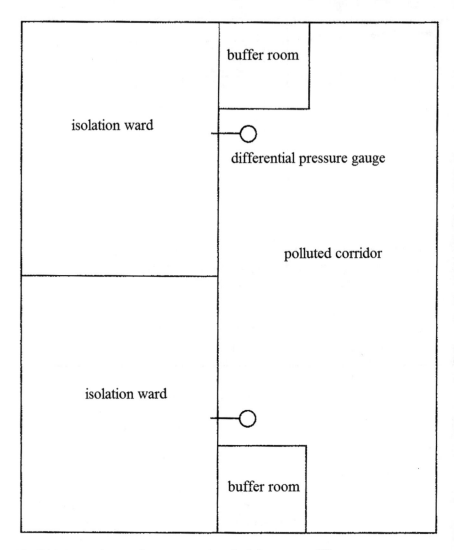

Fig. 7.16 In complete step-by-step expression of relative pressure difference

$$\Delta P_{0,n} = (\pm \Delta P_n) + (\pm \Delta P_{n-1}) + (\pm \Delta P_{n-2}) + \cdots + (\pm \Delta P_1) \qquad (7.2)$$

where $\Delta P_{0,n}$ is the absolute pressure difference at room n. The value of n is the sequence number of the room n along the passage from outdoors towards indoors. ΔP_n is the magnitude of the relative pressure difference between room n and room (n − 1). ΔP_{n-1} is the magnitude of the relative pressure difference between room (n − 1) and room (n − 2), and so on.

Fig. 7.17 Schematic of piezometric tube connecting the atmosphere

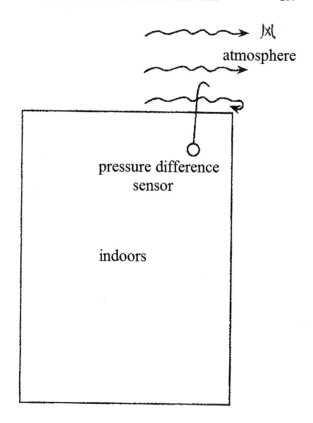

Fig. 7.18 Schematic for design of relative pressure difference in plane

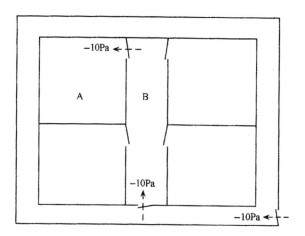

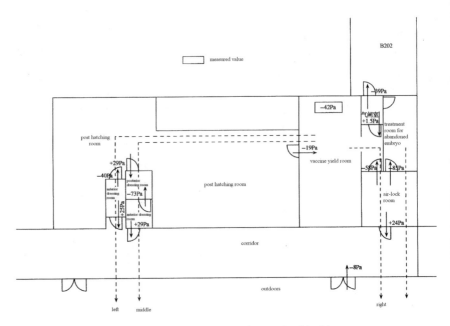

Fig. 7.19 Plane layout of a negative pressure room in an animal health company

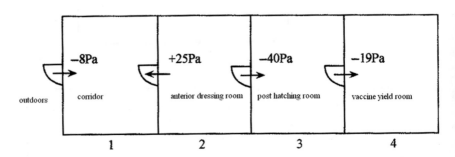

Fig. 7.20 Simplified left route

Next we will take the plane layout of a negative pressure room in an animal health company as an example, which is shown in Fig. 7.19.

Except for the treatment room for abandoned embryo which forms one passage itself, there are three routes for connecting the ambient with the inoculation chamber which needs the highest requirement. These routes are expressed with dashed lines.

(a) The left route can be simplified as Fig. 7.20. We obtain

$$\Delta P_{0,4} = (-\Delta P_4) + (-\Delta P_3) + (+\Delta P_2) + (-\Delta P_1)$$
$$= (-19) + (-40) + (+25) + (-8) = -42\,\text{Pa}$$
$$\Delta P_{0,3} = (-\Delta P_3) + (+\Delta P_2) + (-\Delta P_1) = (-40) + (+25) + (-8) = -23\,\text{Pa}$$

(b) The middle route can be simplified as Fig. 7.21. We obtain

$$\Delta P_{0,5} = (-\Delta P_5) + (+\Delta P_4) + (-\Delta P_3) + (+\Delta P_2) + (-\Delta P_1)$$
$$= (-19) + (+29) + (-73) + (+29) + (-8) = -42\,\text{Pa}$$
$$\Delta P_{0,4} = (+\Delta P_4) + (-\Delta P_3) + (+\Delta P_2) + (-\Delta P_1)$$
$$= (+29) + (-73) + (+29) + (-8) = -23\,\text{Pa}$$

(c) The right route can be simplified as Fig. 7.22. We obtain

$$\Delta P_{0,3} = (-\Delta P_3) + (+\Delta P_2) + (-\Delta P_1) = (-58) + (+24) + (-8)$$
$$= -42\,\text{Pa}$$

It is shown that when anyone route is chosen among the left, middle and right routes, the absolute pressure differences for the inoculation chamber obtained with relative pressure difference are the same. This is the same as the in situ measured absolute pressure difference.

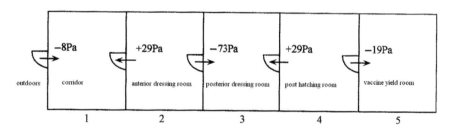

Fig. 7.21 Simplified middle route

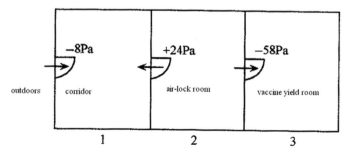

Fig. 7.22 Simplified right route

Similarly the absolute pressure difference in the treatment room for abandoned embryo can be obtained:

$$\Delta P = (-85) + (+24) + (-8) = -62\,\text{Pa}$$

It is known from Fig. 7.21 that even though the relative pressure in the inoculation chamber compared with that in the post hatching room is "+10 Pa", the absolute pressure difference is still:

$$\Delta P = (+10) + (-40) + (+25) + (-8) = -13\,\text{Pa}$$

This means because the relative pressure difference in the post hatching room close to the inoculation chamber is −40 Pa, although the relative difference between the inoculation chamber and the post hatching room reaches +10 Pa, the condition of negative pressure compared with atmospheric pressure can still not be changed.

To determine whether the relative pressure in a room compared with atmospheric pressure is negative or not, calculation with various adjacent rooms connecting outdoors should be performed.

5. Differential pressure gauges should be set at the view height on each exterior wall of the region where the pressure difference is required.

7.7 Design Case

7.7.1 Self-circulation System with Fan Coil Unit

Since the indoor cooling load is dealt with the fan coil unit, the capacity of the outdoor air handling unit reduces.

Because the bacterial removal efficiency of the HEPA filter on the return air grille reaches more than 99.9999%, one bacterium can penetrate through only when the indoor bacterial concentration reaches millions per cubic meter. In common situation, indoor bacterial concentration cannot reach such high value. Therefore, the return air is quite clean. The problem for deposition and accumulation of dust and bacteria on fan coil unit is not a concern.

Therefore, air filters with performance lower than sub-HEPA filter can be installed for the supplied air, such as the low-resistance and high and medium efficiency air filter, or even fine filter. The unit with static pressure between 30 and 50 Pa should be selected.

The problem of condensation should be paid attention to.

The system is shown in Fig. 7.23.

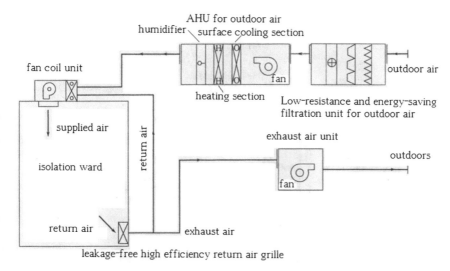

Fig. 7.23 Self-circulation system with fan coil unit

7.7.2   Air Supply Outlet and Fan System

Condensation water is very likely to appear in the coil when the fan coil unit is adopted, which is not expected to occur. So in the current scheme, the coil is abandoned, which is replaced by the ordinary supplied air fan. The return air needs to pass through the outdoor air handling unit, which increases the burden of the outdoor air handling unit.

The system is shown in Fig. 7.24.

7.7.3   Air Supply Outlet and Indoor Self-circulation
Fan System

Because there are 1–2 patients in the ward, the cooling and humidity loads are very small. If there is no need of cooling and humidity handling for the return air, return air will not pass through the outdoor air handling unit. The outdoor air handling unit is only responsible for the cooling and humidity loads from the outdoor air. In this case, there is no problem for formation of condensation water. Liu Hua proposed the specific scheme which is shown in Fig. 7.25.

In this scheme, by reducing the dew point of the machine and increasing the outdoor air volume, the problem of the outdoor air handling unit to treat the entire humidity load is solved. When the outdoor air volume is increased, the air change rate of the outdoor air for the double-bed ward is only 3.1 h^{-1}, which only increases by 1 h^{-1}.

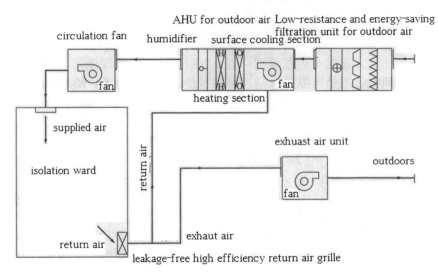

Fig. 7.24 Scheme one with air supply outlet and fan system

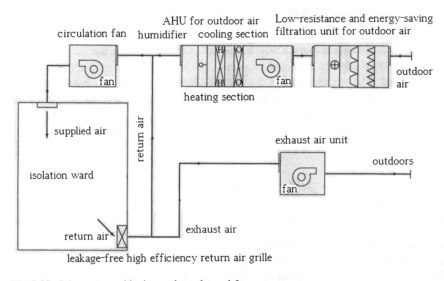

Fig. 7.25 Scheme two with air supply outlet and fan system

Based on the meteorological condition in Shanghai, the surface air cooler with six rows of coils in the outdoor air handling unit can meets the requirement of the design ($t < 26$ °C and $\varphi < 60\%$). When other types of coils are used, eight rows of coils are enough.

The system is shown in Fig. 7.25.

7.7.4 Layout Plan for an Isolation Ward as an Example

Figure 7.26 shows the schematic of the system which was designed by Liu Hua and Liang Lei. One example of the double-bed isolation ward for this scheme is shown in Fig. 7.27.

1. Outline

    The isolation ward is located on the 12th floor with area 1600 m². There are twelve negative pressure isolation wards and four ordinary wards. Every two wards share the same anterior buffer room. Each ward has its posterior buffer room. Moreover, buffer room is placed at all the entrance and exit. This realizes the feature of "Three-Area-Two-Buffer". Three-Area means the clean, the potentially polluted and the polluted areas. Two-Buffer means the buffer room between the ward and the corridor, and the buffer room between the corridor and outdoors. The isolation capability is greatly increased. Independent air conditioning system is used for each area. In the clean area, the scheme with the fan coil unit and the dedicated outdoor air handling unit is used. In the potentially polluted area, the dedicated all air system is used. In the polluted area, independent all air system is adopted for every isolation ward, which operates with two gears of velocity. In other regions, an independent all air system is adopted. The air change rates in the ward and the buffer room are 12 and 60 h⁻¹, respectively.

    The gradient of the pressure is also marked in Fig. 7.26.

2. Simplified analysis
3. Corridors are set in front and back of the ward, which realizes the separation of clean and dirty matter.
4. Buffer rooms are set in front and back of the ward. Buffer rooms are also set between the corridor and outdoors. These greatly increase the isolation

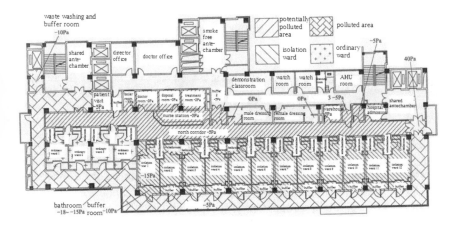

Fig. 7.26 Plane layout of the isolation ward

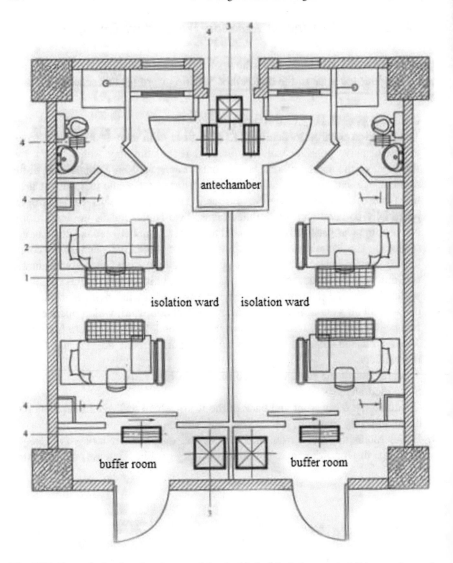

Fig. 7.27 Example for the plane layout of the double-bed isolation ward. *1* Primary air supply grille; *2* Secondary air supply slot; *3* Air supply outlet with sub-HEPA filter; *4* Air return (or exhuast) opening which includes leakage-free high efficiency air return device

performance and safety. By adapted to local conditions, there is no obvious occupation of floor area. Positive pressure should be kept in the buffer room 4.

5. The ward is set at one side, while the auxiliary rooms are set at the other side. The arrangement is uniform, which is good for usage and pollution control.

6. The opening direction for the door of the anterior should be opposite, since the pressure in the anterior is higher than that in the ward.

Table 7.6 Design parameters

Room	Degree of static pressure	Value of static pressure (relative to atmospheric pressure), Pa	Total air change rate, h⁻¹	Outdoor air volume, m³/h	Exhaust air volume, m³/h	Temperature, °C	Relative humidity, %	Noise level, dB(A)	Illuminance level indoors, Lx
Bathroom of the negative pressure isolation ward	—	>15	6–10	-		22–26	40–65	-	50
Negative pressure isolation ward	—	15	8–12	60 m³/h or 3–4 h⁻¹, or full fresh air	All exhaust air	22–26	40–65	≤50	100
Buffer room outside the ward	-	10	≥60	-	self-circulation is allowed	22–26	-	-	50
Interior corridor	-	5	6–10	-	-	20–27	30–65	≤60	100
Office and movable rooms connecting the interior corridor	+	0–5	6–10	40	-	20–27	30–65	≤60	150
Unmanned auxiliary room	0	0	-	-	-	-	-	-	-
Exterior buffer room for the interior corridor	+ or −	5–10	≥60	-	-	20–27	-	-	50
Clean corridor	0	0	-	-	-	-	-	-	-

Note (1) The lower limits for the temperature and humidity are parameters in winter, while the upper limits are parameters in summer; (2) "-" means there is no specific requirement, which needs to be determined by calculation or requirement and then tested; (3) Temperature in the buffer room should be the same as the entrance into the buffer room; (4) When there is room for instrument and device, the temperature and air change rate should be determined by needs

7. Air supply outlet should be placed on the ceiling above the middle of the two beds in the double-bed ward, which is beneficial to control the dispersion of pollutant.
8. Since the leakage-free negative pressure high efficiency air exhaust device is installed at the return air grille, the return air can be switched on. Because the bacterial removal efficiency of the HEPA filter on the return air grille reaches more than 99.9999% and there is no leakage on the frame of air filter, one bacterium in the return air can penetrate through only when the indoor bacterial concentration reaches millions per cubic meter or even tens of millions per cubic meter. But the probability of this incidence is extremely small.
9. For the installation style of the fan coil unit on the ceiling, the problem for the accumulation and removal of condensation water once occurred should be solved.

7.7.5   Design Parameters

Table 7.6 shows the design parameters of the negative pressure isolation ward for reference.

References

1. AIA, *Guidelines for Design and Construction of Hospital and Health Care Facilities* (1998)
2. J. Shen, Y. Liu, Isolation room and HVAC design for SARS rooms. J. HV&AC **33**(4), 10–14 (2003)
3. Z. Xu, Q. Wang, Y. Zhang, H. Liu, W. Niu, R. Wang, The safe distance of a biosafety lab considered from the angle of exhaust air diffusion. Build. Sci. **20**(4), 46–50 (2004)
4. Z. Xu, *Introduction of Cleanroom and Design of its Controlled Environment* (Chemical Industry Press, Beijing, 2008), pp. 248–250

Printed in the United States
By Bookmasters